A BASIC/INTERMEDIATE COURSE
FOR WATER SYSTEM OPERATORS

Volume 4

INTRODUCTION TO Water Quality Analyses

PRINCIPLES and PRACTICES of
WATER SUPPLY OPERATIONS

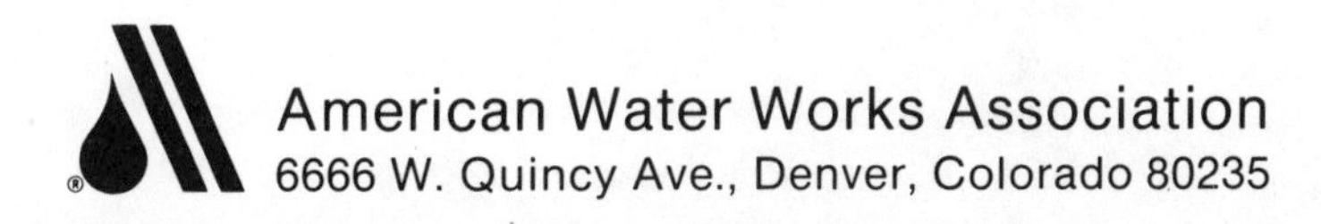
American Water Works Association
6666 W. Quincy Ave., Denver, Colorado 80235

ISBN 0-89867-199-X

Foreword

Introduction to Water Quality Analyses is the fourth volume in a five-part handbook series designed for use in a comprehensive training program entitled "Principles and Practices of Water Supply Operations." Volume 4 provides system operators with practical, working knowledge concerning required and recommended drinking water standards, the importance of monitoring water quality, and the relationship between analyses, unit process control, and the quality of treated water in the distribution system.

Other student volumes in the series include:

Volume 1	*Introduction to Water Sources and Transmission*
Volume 2	*Introduction to Water Treatment*
Volume 3	*Introduction to Water Distribution*
Reference Handbook	*Basic Science Concepts and Applications*

Instructor guide and solutions manuals have been prepared with detailed lesson plan outlines, resource materials, and examination questions and answers for volumes 1 through 4.

Course content in the water supply operations series has been developed to meet training requirements of basic to intermediate grades of certification in water treatment and water distribution system operations. The modular format used throughout the series provides the flexibility needed to conduct both short- and long-term vocational training.

The reference handbook is a required companion textbook that is correlated with volumes 1 through 4 through footnote references. The purpose of a separate reference book is to provide the student with supplementary reading in the areas of mathematics, hydraulics, chemistry, and electricity.

The development of the training materials in the water supply operations series was made possible by funding from the US Environmental Protection Agency, Office of Drinking Water, under Grant Agreement No. T900632-01, awarded to the American Water Works Association.

Disclaimer

Acknowledgments

Publication of this volume was made possible through a grant from the US Environmental Protection Agency, Office of Drinking Water, under Grant Agreement No. T900632-01, as part of a national program strategy for providing a comprehensive curriculum in water treatment plant and water distribution system operations. John B. Mannion, Special Assistant for Communications and Training (currently on an intergovernmental personnel assignment to the staff of the American Water Works Association Research Foundation), represented the Environmental Protection Agency, Office of Drinking Water, as project officer; and Bill D. Haskins, Director of Education, served as project manager for the American Water Works Association.

Special recognition for the work contained in this volume is extended to Jack W. Hoffbuhr, Deputy Director, Water Management Division, USEPA, Region VIII; Thomas E. Braidich, Aquatic Biologist, Drinking Water Branch, USEPA, Region VIII; and Mary Kay Cousin, Associate Technical Editor, American Water Works Association. Appreciation also is expressed to all who gave liberally of their time and expertise in providing technical review of manuscripts. In particular, the following are credited for their participation on the review committee or as an independent volunteer reviewer:

Charles R. Beer, Superintendent of Water Treatment, Denver Water Department, Denver, Colo.

James O. Bryant Jr., Director, Environmental Resources Training Center, Southern Illinois University at Edwardsville

Susan A. Castle, Coordinator for Laboratory Operations, Southern Illinois University at Edwardsville

Jack C. Dice, Quality Control Engineer, Denver Water Department, Denver, Colo.

Monte S. Fryt, Laboratory Director, Department of Public Utilities, Water Division, Colorado Springs, Colo.

James T. Harvey, Superintendent of Production, Water Works, Little Rock, Ark.

William R. Hill, Associate, Camp Dresser & McKee, Inc., Boston, Mass.

Kenneth D. Kerri, Professor of Civil Engineering, School of Engineering, California State University at Sacramento

Jack E. Layne, Director of Engineering, Denver Water Department, Denver, Colo.

Ralph W. Leidholdt, Senior Project Engineer, KKBNA, Inc., Denver, Colo.

Michael J. McGuire, Research Engineer, Water Quality, The Metropolitan Water District of Southern California, Los Angeles, Calif.

Andrew J. Piatek Jr., Chief Operator & Superintendent, Borough of Sayreville, N.J.

John F. Rieman, Senior Technical Editor, American Water Works Association, Denver, Colo.

Robert L. Wubbena, President, Economic & Engineering Services, Inc., Olympia, Wash.

Introduction for the Student

Water quality analyses form the foundation of successful water supply operations. The results of these laboratory analyses extend an operator's vision and measurement ability into the worlds of microbiology and inorganic and organic chemistry. The sampling and testing involved are a demanding part of the operator's day-to-day responsibilities, requiring the skill to perform a variety of detailed procedures with specialized equipment, a knowledge of the measurements and principles used by chemists and biologists, and an understanding of the laws and regulations that govern the production and distribution of drinking water.

Routine monitoring of water quality is necessary to ensure that each component of the water system (the water source, treatment plant, and distribution system) is in proper operating condition, providing safe, palatable drinking water. Equally important, laboratory analyses allow the operator to monitor the efficiency of various water treatment processes, with laboratory test results forming the basis for process modifications that will increase the economy of operation and improve treated water quality.

To help the operator develop a practical, working knowledge of water quality analyses, this text covers five basic areas:

- Drinking water standards
- Sample collection, preservation, and storage
- Use of laboratory equipment
- Microbiological tests
- Physical/chemical tests.

The two modules on testing describe the significance of each test, the methods of sampling and testing, and perhaps most important, the interpretation of test results. Complete information on equipment needed, reagents needed, and detailed test procedures required to actually conduct each test can be found in either of the following standard references:

- *Standard Methods for the Examination of Water and Wastewater,* latest edition, published jointly by APHA, AWWA, and WPCF.
- *Methods for Chemical Analyses of Water and Wastes,* EPA-625/6-74-003. USEPA, Office of Technology Transfer (1974).

Publications that cover laboratory analysis procedures on a simplified basis include:

- *Simplified Procedures for Water Examination,* AWWA Manual M12, AWWA, Denver, Colo. (1975).
- *Manual of Instruction for Water Treatment Plant Operators,* New York State Department of Health, Albany, N.Y. (No date).

In addition, several laboratory equipment manufacturers and suppliers have prepared handbooks that outline in detail the required equipment, reagents, and, in some cases, the test procedures.

The operator should obtain one of the simplified laboratory publications and have access to either *Standard Methods* or *Methods for Chemical Analysis of Water and Wastes.*

Table of Contents

Appendices

Water Quality Analyses

Module 1

Drinking Water Standards

In 1974, the US Congress passed the SAFE DRINKING WATER ACT* (the SDWA, Public Law 93-523) establishing a cooperative program among local, state, and federal agencies to help ensure safe drinking water in the United States. Under the SDWA, the primary role of the federal government is to develop national drinking water regulations that will protect public health and welfare. The states are assigned the responsibility of implementing the regulations and monitoring the performance of public water systems. The public water systems themselves are responsible for treating and testing drinking water to ensure that its quality consistently meets the standards set by the regulations.

As directed by the SDWA, the US Environmental Protection Agency (USEPA) developed primary and secondary drinking water regulations designed to protect public health and welfare. Table 1-1 shows the regulations that have been developed and their effective dates.

The NATIONAL INTERIM PRIMARY DRINKING WATER REGULATIONS (NIPDWRs, or IPRs) and subsequent amendments cover contaminants that have adverse effects on human health. These regulations are enforceable by the USEPA.

The SECONDARY DRINKING WATER REGULATIONS cover contaminants that adversely affect the aesthetic quality of drinking water, such as taste, odor, and appearance. These regulations are intended as guidelines and are not enforceable by USEPA; however, individual states may choose to enforce some or all of the Secondary Regulations. Although the contaminants covered by the Secondary Regulations do not normally affect health directly, the effects of those contaminants may influence consumers to use water that is aesthetically more

*Words set in SMALL CAPITAL LETTERS are glossary terms. Definitions for these terms can be found in the glossary at the end of this volume.

pleasing—but possibly less safe—than the public supply. In addition, they can increase the costs of operation and maintenance.

The purpose of this module is to explain the national drinking water regulations, both primary and secondary. Most states have adopted regulations as strict as the NIPDWRs. However, since some states have regulations that are more strict, operators should also familiarize themselves with the drinking water regulations in their own state.

After completing this module you should be able to

- Describe the purpose of the National Interim Primary and Secondary Drinking Water Regulations.
- Define MCL and explain what it means.
- List the contaminants covered by both the primary and secondary regulations and discuss their importance.
- Describe the importance of adequate sampling, testing, record keeping, and reporting.

Table 1-1. National Drinking Water Regulations

Regulation	*Promulgation Date**	*Effective Date*	*Primary Coverage*
National Interim Primary Drinking Water Regulations	12/24/75	6/24/77	Inorganic, organic, and microbiological contaminants and turbidity
1st Amendment	7/9/76	6/24/77	Radionuclides
2nd Amendment	11/29/79	Varies, depending on system size†	Trihalomethanes
3rd Amendment	8/27/80	2/27/82	Special monitoring requirements for corrosion and sodium
National Secondary Drinking Water Regulations	7/19/79	7/19/79	Guidelines for contaminants affecting color, taste, and appearance of water

*The date that the regulation was published in final form in the Federal Register.
†See Table 1-4.

1-1. The National Interim Primary Drinking Water Regulations

The NIPDWRs are the first drinking water regulations to apply to all public water systems in the United States. After detailed studies of the health effects of various levels of contaminants in drinking water are completed, Revised Primary Drinking Water Regulations will be developed and published. In order to provide adequate public health protection until the revised regulations are developed, the NIPDWRs have been amended to include additional contaminants determined to have adverse effects on human health. The national regulations and amendments are summarized in Table 1-1.

The Safe Drinking Water Act makes it clear that the owners and operators of public water systems are responsible for ensuring that their systems meet the regulations. To help fulfill this responsibility, operators need to be familiar with the following five areas of the SDWA and the NIPDWRs:

- Definition of "public water system"
- Maximum contaminant levels (MCLs)
- Sampling frequencies
- Record-keeping requirements
- Reporting requirements.

Each of the above items is discussed in this module. The discussion refers only to the USEPA regulations; if a state has adopted its own regulations, they could be more strict.

Definition of Public Water Supply

The NIPDWRs apply to all PUBLIC WATER SYSTEMS. Public water systems are defined in the SDWA as those systems that either (1) have 15 or more service connections or (2) regularly serve an average of 25 or more people daily for at least 60 days each year.

There are two types of public water systems defined by the NIPDWRs: community systems and non-community systems (Figure 1-1). A COMMUNITY SYSTEM is one that serves a residential (year-round) population—a population in which the same people drink the water from the same system regularly over a long period of time. A NON-COMMUNITY SYSTEM is one that serves intermittent users, such as tourists. For example, a campground may serve hundreds of people daily; but they are probably different people each day, so no individual drinks very much of the campground's water. Since certain contaminants have adverse health effects only when consumed regularly over a long period of time, the distinction between community and non-community systems is important in determining which contaminants must be monitored to protect public health.

Maximum Contaminant Levels and Health Effects

Under the SDWA regulations, a MAXIMUM CONTAMINANT LEVEL, or MCL is the highest allowable concentration of a particular contaminant in drinking water.

Community Water Systems provide water to year-round residents.

Non-Community Water Systems serve travelers and intermittent users.

Figure 1-1. Two Types of Public Water Systems

An MCL is usually expressed in MILLIGRAMS PER LITRE (mg/L), which is the same for the purposes of water quality analysis as PARTS PER MILLION (ppm).

Table 1-2 lists the MCLs for the contaminants included in the NIPDWRs. Table 1-3 summarizes the adverse health effects of those contaminants. Except for those relating to nitrate, microbiological contaminants, and turbidity, the MCLs are designed to prevent health effects caused by drinking water from the same system over a long period of time and are only applicable to community water systems. Nitrate, bacteria, and turbidity, on the other hand, can be related to rapid onset of health problems, so non-community as well as community systems are required to meet the MCLs for those contaminants. The following paragraphs discuss briefly the contaminants and MCLs shown in Table 1-2.

Inorganic chemicals. INORGANIC CHEMICALS covered by the NIPDWRs include fluoride and nitrate, as well as several naturally occurring elements. Arsenic, barium, selenium, and fluoride can be found naturally in drinking water sources, particularly ground water. The presence of chromium, mercury, silver, and nitrate in drinking water is usually due to man-made contamination. Lead and cadmium are generally found as a result of corrosion of piping materials in the distribution and household plumbing systems.

Fluoride has a range of MCLs depending on air temperature. The range was established to compensate for the higher consumption of water and consequent increased fluoride intake in warmer climates. The range also provides more flexibility than a single MCL in allowing the beneficial effects of natural fluoride to be gained with an adequate safety margin. The MCLs should be applied only to water systems with naturally occurring fluorides. When adding a fluoride compound to increase the fluoride concentration, follow the recommendation of the state health department. Normally, the level that should be maintained is half of the MCL for the local temperature range.

Organic chemicals. The ORGANIC CHEMICALS for which MCLs have been established include various HERBICIDES and INSECTICIDES, as well as TRIHALOMETHANES (THMs). Trihalomethanes are a group of compounds formed when chlorine reacts with humic and fulvic acids, natural organic compounds that occur in decaying vegetation.

Turbidity. For many water systems using surface sources, the TURBIDITY MCL is the most difficult to meet, especially where chemical coagulation and filtration are not performed. Turbidity is caused by small, suspended particles in the water. These particles can be either organic material, such as leaf mold and animal waste, or inorganic material, such as sand and clay. Excessive turbidity is a problem for several reasons.

- It protects microorganisms from chlorine and other disinfectants.
- It acts as a food source for microorganisms, allowing them to survive and multiply in the distribution system.
- It interferes with the maintenance of a chlorine residual in the distribution system.
- It interferes with the test for coliform bacteria.

Turbidity is tested by measuring the amount of light scattered by particles in the water. As the number of particles increases, more light is scattered and a higher turbidity reading is obtained. The measuring instrument is called a NEPHELOMETER, and the readings are expressed as NEPHELOMETRIC TURBIDITY UNITS (NTU) or turbidity units. From a public health standpoint, turbidity in drinking water should be less than 1 NTU.

Microbiological contaminants. The microbiological MCLs are based on COLIFORM BACTERIA tests, which have been used for many years as an indication of the bacteriological quality of water. An indicator test is used since it would be impossible to test rapidly and economically for every PATHOGEN that might be present.

Table 1-2. Maximum Contaminant Levels

Type of Contaminant (Community systems to which MCL applies)	*Type of Contaminant (Non-community systems to which MCL applies)*	*Maximum Contaminant Levels (MCLs)*		
Inorganic chemicals (systems using surface or ground water)	Inorganic chemicals: other than nitrate (state option)	Arsenic	0.05	mg/L
		Barium	1	mg/L
		Cadmium	0.010	mg/L
		Chromium	0.05	mg/L
		Lead	0.05	mg/L
		Mercury	0.002	mg/L
		Selenium	0.01	mg/L
		Silver	0.05	mg/L
		Fluoride (annual average of maximum daily air temperatures)		
		a) 53.7°F & below	2.4	mg/L
		b) 53.8–58.3°F	2.2	mg/L
		c) 58.4–63.8°F	2.0	mg/L
		d) 63.9–70.6°F	1.8	mg/L
		e) 70.7–79.2°F	1.6	mg/L
		f) 79.3–90.0°F	1.4	mg/L
	Inorganic chemicals: nitrate* (systems using surface or ground water)	Nitrate (as N)	10	mg/L
Organic chemicals: pesticides, herbicides (surface-water systems only)	Organic chemicals: pesticides, herbicides (state option)	Endrin	0.0002	mg/L
		Lindane	0.004	mg/L
		Methoxychlor	0.1	mg/L
		Toxaphene	0.005	mg/L
		2,4-D	0.1	mg/L
		2,4,5-TP (Silvex)	0.01	mg/L
Organic chemicals: THMs (all systems that add a disinfectant [oxidant] to the water and serve populations greater than 10,000)	Organic chemicals: THMs (state option)	THMs	0.1	mg/L
Turbidity (surface-water systems only)	Turbidity (surface-water systems only)	Monthly average may not exceed 1 NTU (5 NTU monthly average may apply at state option)		
		Average of any 2 consecutive days may not exceed 5 NTU		

*For all non-community water systems, initial sampling and testing must be conducted for nitrates. Routine sampling and testing, however, is at state option.

Table 1-2. Maximum Contaminant Levels *(continued)*

Type of Contaminant (Community systems to which MCL applies)	*Type of Contaminant (Non-community systems to which MCL applies)*	*Maximum Contaminant Levels (MCLs)*	
Microbiological contaminants (systems using surface or ground water)	Microbiological contaminants (systems using surface or ground water)	*When using membrane filter test:* 1 colony/100 mL for the average of all monthly samples. 4 colonies/100 mL in more than 1 sample if less than 20 samples are collected per month. 4 colonies/100 mL in more than 5 percent of the samples if 20 or more samples are examined per month. *When using multiple-tube fermentation test (10-mL portions):* Coliform shall not be present in more than 10 percent of the portions per month. Not more than 1 sample may have 3 or more portions positive if less than 20 samples are examined per month. Not more than 5 percent of the samples may have 3 or more portions positive when 20 or more samples are examined per month.	
Radiological contaminants: natural (systems using surface or ground water)	Radiological contaminants: natural (state option)	Gross Alpha Combined Radium-226 and Radium-228	15 pCi/L 5 pCi/L
Radiological contaminants: man-made (surface-water systems serving populations greater than 100,000)	Radiological contaminants: man-made (state option)	Gross Beta Tritium Strontium-90	50 pCi/L 20,000 pCi/L 8 pCi/L

Table 1-3. Health Effects of Contaminants Regulated by the NIPDWRs

Contaminant	*Health Effect*
Microbiological organisms	Cause various illnesses, some potentially fatal. Common waterborne diseases caused by microorganisms include gastroenteritis, typhoid, bacillary dysentery, infectious hepatitis, amebic dysentery, and giardiasis.
Turbidity	Protects microorganisms from chlorine and other disinfectants, acts as a food source for microorganisms, interferes with maintenance of a chlorine residual in the distribution system, and interferes with coliform testing.
Arsenic	Causes small sores on hands and feet, possibly developing into cancers.
Barium	Causes increased blood pressure and nerve block.
Cadmium	Concentrates in liver, kidneys, pancreas, and thyroid; hypertension is a suspected health effect.
Chromium	Causes skin sensitization, kidney damage.
Lead	Causes constipation; loss of appetite; anemia; tenderness; pain and gradual paralysis in the muscles, especially the arms.
Mercury	Causes inflammation of the mouth and gums; swelling of the salivary glands; loosening of the teeth.
Selenium	Causes staining of fingers, teeth and hair; general weakness; depression; irritation of the nose and throat.
Silver	Causes permanent gray discoloration of skin, eyes, and mucous membranes.
Fluoride	Causes stained spots on teeth (mottling)—the amount of discoloration depends on the amount of fluoride ingested.
Nitrate	Causes temporary blood disorder in infants—can be fatal.
Pesticides Endrine Lindane Methoxychlor Toxaphene	Cause symptoms of poisoning which differ in intensity. The severity is related to the concentration of these chemicals in the nervous system, primarily the brain. Mild exposure causes headaches, dizziness, numbness and weakness of the extremities. Severe exposure leads to spasms involving entire muscle groups, leading in some cases to convulsions. Suspected of being carcinogenic.
Herbicides 2,4-D 2,4,5-TP	Cause liver damage and gastrointestinal irritation.
Trihalomethanes	Suspected as possible carcinogens.

*The health effects of microbiological organisms, turbidity, and nitrate result from short-term exposure. All other health effects result from exposure over a long period of time.

Radiological contaminants. The radiological MCLs are complex, but most water systems will only have to monitor for GROSS ALPHA ACTIVITY. The NIPDWRs provide that when the gross alpha activity is found to be less than 5 PICOCURIES per litre, no further testing for natural radioactivity is required. In addition, only those water systems using surface-water sources and serving more than 100,000 people have to monitor for the man-made RADIONUCLIDES.

1-2. Monitoring and Reporting Requirements

To ensure that drinking water meets the standards set by the SDWA, utilities are required to perform regular sampling and testing of the water they supply to consumers. The NIPDWRs specify sampling frequency and location, testing procedures, and requirements for record keeping and routine reporting to the state or USEPA. The regulations also cover special reporting procedures to be followed if a contaminant is found to exceed an MCL.

A working knowledge of the sampling, testing, and record-keeping requirements is especially important to the operator; these procedures are part of the daily operating routine. The following material summarizes the requirements of the NIPDWRs; in some cases, individual states may have stricter requirements.

Monitoring Frequencies

The NIPDWRs specify minimum MONITORING frequencies and require that all analyses except turbidity be conducted in laboratories certified by USEPA or the state. An important responsibility of the operator is to ensure that the proper number of samples are taken carefully and sent to the laboratory on time.

Table 1-4 shows the required sampling and testing frequencies for community systems. Monitoring required for ground-water sources may be less frequent than for surface water since ground water is generally more uniform in quality and less subject to contamination. Table 1-5 shows the monitoring requirements for sodium and corrosion. Although there are no MCLs for these characteristics, monitoring is required because of their public health importance. As with all contaminants, a state can require more frequent monitoring.

Most water systems will not be able to afford their own labs and will instead send ROUTINE SAMPLES to state or commercial laboratories for analysis. However, all systems using surface sources should purchase a nephelometer to use in the daily turbidity tests.

An important phase of monitoring is the CHECK SAMPLING required when analysis of a sample shows it to have a contaminant concentration that exceeds an MCL. Check sampling provides a safeguard against sampling or laboratory error. Check-sampling requirements are summarized in Table 1-6.

Record Keeping

Data developed through the water-quality testing required under the SDWA can be useful in evaluating system performance, planning improvements, and

Table 1-4. Required Sampling

What Tests	*Where Samples Taken*	*How Often (Community system)*	*How Often (Non-community system)*
Inorganics	At the consumer's faucet*	Systems using surface water: every year Systems using ground water only: every 3 years	State option except for nitrate†
Organics: except THMs	At the consumer's faucet*	Systems using surface water: every 3 years Systems using ground water only: state option	State option
Organics:THMs	25 percent at extremes of distribution system; 75 percent at locations representative of population distribution	Systems serving populations of 10,000 or more: 4 samples per quarter per plant §	State option
Turbidity	At the point(s) where water enters the distribution system	Systems using surface water: daily Systems using ground water only: state option	Systems using surface or surface and ground water only: daily Systems using ground water only: state option
Coliform bacteria	At the consumer's faucet*	Depends on number of people served by the water system (See Appendix A.)	Systems using surface and/or ground water: 1 per quarter (for each quarter water is served to public)

*The faucets selected must be representative of conditions within the distribution system.

†Although routine nitrate monitoring is established at state option, the initial monitoring is required and should have been completed by June 1979.

§Systems using multiple wells drawing raw water from a single aquifer may, with state approval, be considered one treatment plant for determining the required number of samples.

Table 1-4. Required Sampling *(continued)*

What Tests	*Where Samples Taken*	*How Often (Community system)*	*How Often (Non-community system)*
Radiochemicals: natural	At the consumer's faucet*	Systems using surface water: every 4 years Systems using ground water only: every 4 years	State option
Radiochemicals: man-made	At the consumer's faucet* (at state option)	System using surface water serving populations greater than 100,000: every 4 years All other systems: state option	System using surface and/or ground water: state option

*The faucets selected must be representative of conditions within the distribution system.

writing annual reports. Table 1-7 shows how long the NIPDWRs require test results and related records to be kept.

Summaries of laboratory reports of bacteriological or chemical analyses may be kept instead of the individual lab reports. Any summary, however, must contain the information shown in Table 1-8.

Reporting to the State

To ensure that prompt attention is given to potential public health problems, the NIPDWRs require water systems to submit routine reports to the appropriate regulatory agency. Generally, reports will go to the state water supply agency. However, in some states the USEPA has the primary responsibility for implementing the Safe Drinking Water Act, and in those states the reports will go either directly to the USEPA regional office or through the state to USEPA. The water systems in those states will be notified of the procedures to be used. Wherever the "state" is mentioned in the following discussion, it should be understood to mean the state or USEPA, whichever is applicable.

There are three types of reports that must be sent to the state:

- Routine sample reports
- Check sample reports
- Violation reports.

Routine sample reports. If a state laboratory is doing the testing, routine sample results need not be reported by the water system, since the state will

Table 1-5. Special Monitoring Requirements for Sodium and Corrosion (community systems only)*

Test	*Where Sample Taken*	*How Often*
Sodium	Entry point to distribution system for each source	Systems using surface water: annually Systems using ground water only: every 3 years
Corrosivity includes those characteristics known to indicate corrosivity: pH Calcium hardness Total dissolved solids (TDS) Temperature Langelier Index	Entry point to distribution system for each source	Once unless additional monitoring required by state or EPA

*First analyses must be completed by February, 1983.

Table 1-6. Check-Sampling Requirements

Type of MCL Exceeded	*Check Samples**
Inorganic (except nitrate)	3 within 1 month
Nitrate	1 within 24 hours
Organic (except THMs)	3 within 1 month
THMs	Continue quarterly monitoring
Radiological	
Natural	1 per quarter until annual average concentration no longer exceeds MCL
Man-made	1 per month until concentration no longer exceeds MCL
Turbidity	1 within 1 hour
Bacteriological	
Membrane filter	1 per day until 2 consecutive samples show less than 1 coliform/100 mL
Multiple-tube fermentation	1 per day until 2 consecutive samples show no positive tubes

*After results from initial or routine sample are obtained.

Table 1-7. Record-Keeping Requirements

Records Pertaining to	*Time Period*
Bacteriological and turbidity analyses	5 years
Chemical analyses	10 years
Actions taken to correct violations	3 years
Sanitary survey reports	10 years
Exemptions	5 years following expiration

Table 1-8. Lab Report Summary Requirements

Sampling Information	*Analysis Information*
Date, place, and time of sampling	Date of analysis
Name of sample collector	Laboratory conducting analysis
Identification of sample	Name of person responsible for analysis
Routine or check sample	Analytical method used
Raw or treated water	Analysis results

already have the results. However, if a commercial laboratory or the system's own laboratory is doing the testing, the results of any test required by the NIPDWRs must be reported within the first 10 days following the month in which the results are received. This allows all data for each month to be summarized and sent in at one time, instead of sending in each individual test result immediately after it is received. For example, water supplies using surface-water sources will have to report their daily turbidity readings to the state at the end of each month.

If the results indicate that an MCL for one of the inorganic (except nitrate) or organic contaminants has been exceeded, the state must be notified within seven days and the check sampling procedure must be started to be completed within the time shown in Table 1-6. Because high nitrate concentrations can be an acute hazard to infants, the report must be made and required check sampling performed within 24 hours.

Check sample reports. The check sampling required by the NIPDWRs (Table 1-6) is performed to determine whether a problem still exists, as well as to provide a safety factor against sampling or laboratory error. In most cases, the reporting of check samples is done monthly, as for routine samples. The exceptions are shown in Table 1-9.

Violation reports. A violation of the NIPDWRs must be reported to the state within 48 hours. Violating an MCL, failing to monitor according to the required schedule, and failing to use the proper test procedures are all violations that must be reported. Table 1-10 indicates what constitutes a violation of the various MCLs.

Table 1-9. Reporting Requirements for Check Sampling

Contaminant	*Check-Sample Reporting*
Microbiological	Must report to state within 48 hours when any check sample confirms the presence of coliform bacteria.
Turbidity	Must report to state within 48 hours if check sample confirms MCL has been exceeded.
Nitrate	Must report to state within 24 hours if check sampling confirms MCL has been exceeded.
All others	Must be reported to the state within 10 days after the end of the month in which the sample was received.

Table 1-10. MCL Violations

Contaminant	*Violation*
Inorganic chemicals (except nitrate) and organic chemicals (except THMs)	If average of results from initial sample plus 3 check samples exceeds MCL.
Nitrate	If average of results from initial sample plus the check sample exceeds MCL.
THMs	If average of results from present quarter plus those of 3 preceding quarters exceeds MCL.*
Radionuclides (natural and man-made)	If average annual concentration exceeds MCL.†
Turbidity	a. If monthly average of daily readings exceeds 1 NTU (or 5 NTU if allowed by state). b. If average of 2 consecutive daily readings exceeds 5 NTU.
Microbiological (coliform testing): membrane filter and multiple-tube fermentation	If any of the MCLs are exceeded.

*Quarter means a 3-month period. For convenience, calendar quarters are used.
†Based on individual analyses of 4 consecutive quarterly samples or a single analysis of an annual composite of 4 quarterly samples.

Public Notification

One of the most important sections of the Safe Drinking Water Act is the requirement that water suppliers notify their customers when their water systems are in violation of the NIPDWRs. The purpose of the notice is to protect consumers from water that may be temporarily unsafe, as well as to increase public awareness of the problems water systems face and the costs of supplying safe drinking water. Table 1-11 indicates the public notification requirements for different violations. Notification by mail is sufficient for all cases except violation of an MCL, which requires newspaper and broadcast notice as well.

Table 1-11. Public Notification Requirements

Violation or Condition	*Required Notification*		
	Mail	*Newspaper*	*Broadcast*
Violation of an MCL	X	X	X
Failure to monitor	X		
Failure to follow compliance schedule	X		
Failure to use approved testing procedure	X		
System granted a variance or exemption	X		

Although the NIPDWRs do not give detailed procedures for writing a public notice, they do state that the notice:

- Should be conspicuous and present the facts concerning the violation
- Should not use overly technical language or small print
- Should explain the public health significance of the violation
- Should explain what the system is doing to correct the problem.

Further information on how to write public notices is included in the sugggested references.

Mail notice. When notice by mail is required, the consumer must be notified in the next water bill or by special mailing, but in any case within three months of the violation or granting of a variance or exemption. This notice must be repeated every three months as long as the condition exists.

Newspaper notice. When newspaper notice is required, the notice must be published on three consecutive days in a newspaper serving the area. Publication must be completed within 14 days after the water supplier learns of the violation. If only a weekly paper serves the area, the notice must appear in three consecutive issues of the paper. If there are no newspapers serving the area, the notice must be posted in the local post office.

Broadcast notice. When broadcast notice is required, a copy of the notice must be provided to radio and TV stations serving the area within seven days after the water supplier learns of the MCL violation.

1-3. The Secondary Drinking Water Regulations

The Secondary Drinking Water Regulations cover contaminants that affect the taste, odor, or appearance of drinking water. Unlike the NIPDWRs, the Secondary Regulations are only guidelines; they are not federally enforceable. However, many states have chosen to enforce some or all of the MCLs as shown in Table 1-12. The MCLs represent reasonable goals for drinking water quality.

Table 1-12. Secondary Maximum Contaminant Levels

Contaminant	*Level*
Chloride	250 mg/L
Color	15 Color Units (CU)
Copper	1 mg/L
Corrosivity	Non-corrosive
Foaming Agents	0.5 mg/L
Iron	0.3 mg/L
Manganese	0.05 mg/L
Odor	3 Threshold Odor Number (TON)
pH	6.5–8.5
Sulfate	250 mg/L
Total Dissolved Solids (TDS)	500 mg/L
Zinc	5 mg/L

The states may establish higher or lower levels which may be appropriate to local conditions, provided that public health and welfare are not adversely affected.

Although established only as guidelines, the Secondary Regulations are important since the contaminants they cover affect the characteristics by which consumers judge their water: how it looks, tastes, and smells. Where the public drinking water is colored, smells like rotten eggs, or tastes medicinal, consumers will often find other sources of water which may be less safe. Table 1-13 indicates some of the adverse effects of secondary contaminants.

Monitoring of the secondary contaminants is not required, but the results of routine testing can be useful to the operator. A significant change in test results from one test to the next could indicate operating problems or contamination.

Table 1-13. Adverse Effects of Secondary Contaminants

Contaminant	*Adverse Effect*	*Contaminant*	*Adverse Effect*
Chloride	Causes taste. Adds to TDS and scale. Indicates contamination.	Iron	Discolors laundry brown. Changes taste of water, tea, coffee, and other beverages.
Color	Indicates dissolved organics may be present, which may lead to trihalomethane formation. Unappealing appearance.	Manganese	Discolors laundry. Changes taste of water, tea, coffee, and other beverages.
Copper	Undesirable metallic taste.	Odor	Unappealing to drink. May indicate contamination.
Corrosivity	Corrosion products unappealing to consumers. Causes taste and odor. Corrosion products can affect health. Corrosion causes costly deterioration of water system.	pH	Below 6.5 water is corrosive. Above 8.5 water will form scale, taste bitter.
		Sulfate	Has laxative effect.
		Total Dissolved Solids (TDS)	Associated with taste, scale, corrosion, and hardness.
Foaming Agents	Unappealing appearance. Indicates possible contamination.	Zinc	Undesirable taste. Milky appearance
Hydrogen Sulfide	Offensive odor. Causes black stains on contact with iron. Can accumulate to deadly concentration in poorly ventilated areas. Flammable and explosive.		

Selected Supplementary Readings

EPA Secondary Regulations Promulgated. *OpFlow*, 5:10:1 (Oct. 1979).

Ferguson, Susan. Sampling and Testing: Essential for Quality Control. *OpFlow*, 7:9:5 (Sept. 1981).

Hoffbuhr, Jack & Chaussee, Dean. Are You Ready for the Safe Drinking Water Act?—Part I. *OpFlow*, 3:4:1 (Apr. 1977).

Hoffbuhr, Jack & Chaussee, Dean. The Safe Drinking Water Act Goes Into Effect Next Month—Are You Prepared?—Part II. *OpFlow*, 3:5:1 (May 1977).

How the Laboratory Can Assist the Treatment Plant Operator—Part I. *OpFlow*, 5:4:4 (Apr. 1979).

How the Laboratory Can Assist the Treatment Plant Operator—Part II. *OpFlow*, 5:5:4 (May 1979).

Manual of Instruction for Water Treatment Plant Operators. New York State Dept. of Health. Albany, N.Y. (1975).

National Interim Primary Drinking Water Regulations. Office of Water Supply (OWS), USEPA, Washington, D.C. EPA-570/9-76-003. (1976).

National Secondary Drinking Water Regulations. Office of Drinking Water (ODW), USEPA, Washington, D.C. EPA-570/9-76-000. (July 1979).

SDWA, IPRs, and MCLs Simplified. *OpFlow*, 3:7:1 (July 1977).

SDWA, IPRs, and MCLs Simplified. *OpFlow*, 3:8:6 (Aug. 1977).

SDWA, IPRs, and MCLs Simplified. *OpFlow*, 3:9:1 (Sept. 1977).

SDWA, IPRs, and MCLs Simplified. *OpFlow*, 4:2:1 (Feb. 1978).

The Safe Drinking Water Act Self-Study Handbook for Community Water Systems. AWWA, Denver, Colo. (1978).

Glossary Terms Introduced in Module 1

(Terms are defined in the Glossary at the back of the book.)

Check sampling
Coliform bacteria
Community system
Gross alpha activity
Herbicide
Inorganic chemical
Insecticide
MCL
Maximum contaminant level
Milligrams per litre (mg/L)
Monitoring
National Interim Primary Drinking Water Regulations
Nephelometer
Nephelometric turbidity unit (NTU)
Non-community system
Organic chemical
Parts per million (ppm)
Pathogen
Picocurie
Public water system
Radionuclide
Routine (required) sample
Safe Drinking Water Act
Secondary Drinking Water Regulations
Trihalomethanes (THMs)
Turbidity

Review Questions

(Answers to Review Questions are given at the back of the book.)

1. When sampling and testing for water quality analysis what three things are required of the water treatment operator?

2. List two benefits of routine monitoring of water quality.

3. What is the responsibility of federal agencies under the Safe Drinking Water Act?

4. What are the responsibilities of state agencies under the Safe Drinking Water Act?

5. What are two types of public water systems described in the NIPDWRs?

6. Which of these two types would be subject to MCLs designed to prevent health effects caused by drinking the water over a long period of time?

7. List five types of contaminants for which MCLs are provided.

8. As a rule, which type of water supply must be monitored more frequently, ground water or surface water? Why?

9. Are the monitoring requirements for all systems the same as those set forth in the NIPDWRs?

10. Describe the actions which must be taken if an MCL for a substance other than nitrate is exceeded.

11. Given the following results (multiple-tube fermentation), determine whether an MCL violation has occurred.

Routine Monthly Sample Analysis

No. of Tubes	*No. of Positive Tubes*
5	0
5	2
5	3
5	0
5	1
5	0

12. Do the following results (multiple-tube fermentation) indicate an MCL violation?

	Routine Sample		*Check Sample*	
Date	*No. of Tubes*	*No. of Positive Tubes*	*No. of Tubes*	*No. of Positive Tubes*
4/1/78	5	0		
4/5/78	5	1		
4/10/78	5	0		
4/14/78	5	3		
4/15/78			5	0
4/16/78			5	0
4/24/78	5	0		
4/27/78	5	0		
4/30/78	5	0		

13. For the monthly results (membrane filter) given below, would the public have to be notified because of any of the results?

Date	*Routine Sample Results*	*Check Sample Results*
4/1/78	0	
4/2/78	1	
4/3/78	1	
4/10/78	0	
4/12/78	5	
4/13/78		2
4/14/78		0
4/15/78	3	0
4/16/78	1	
4/20/78	2	
4/24/78	0	
4/26/78	0	
4/30/78	1	

14. What factor determines the number of coliform analyses that a utility must run under the SDWA?

15. What types of contaminants do the Secondary Drinking Water Regulations cover?

16. Why are the guidelines set forth in the secondary regulations important?

Study Problems and Exercises

1. You have recently been hired as chief operator for a city with a population of 27,500. The water system consists of a surface-water source and several wells pumping from a common aquifer. Design a monitoring program for the city to meet the requirements of the NIPDWRs.

2. Compare your system's monitoring program with the NIPDWRs or state regulations and discuss any changes needed for compliance. What else do you feel should be monitored?

Module 2

Sample Collection, Preservation, and Storage

The first step in water quality analysis is the collection of samples that accurately represent the quality of the water being sampled. This module discusses the types of samples that can be collected, the sample volumes required, and the selection of sample points. General techniques for sample collection, preservation, and storage are also covered. If special sampling procedures are required when testing for a particular constituent, those procedures are discussed in Module 4, Microbiological Tests, or Module 5, Physical/Chemical Tests, along with the procedure for performing the test.

After completing this module you should be able to

- Describe the importance of representative sampling.
- Identify the two general types of samples that can be collected and select the proper type for various sampling and testing situations.
- Select proper sample volumes.
- Establish representative sampling points.
- Collect representative samples.
- Understand the importance and limitations of sample preservation and storage.
- Prepare complete records for collection, preservation, and storage of samples.

2-1. Representative Sampling and Types of Samples

One of the most common causes of error in water quality analysis is improper sampling. When a sample is tested, the test results show only what is in the sample. If those test results are to be useful, the sample must contain essentially the same constituents as the body of water from which it was taken—that is, it must be a REPRESENTATIVE SAMPLE.

There are two general types of samples:

- Grab samples
- Composite samples.

A properly taken grab sample is representative of the quality of the water at the exact time and place the sample was taken. A composite sample is representative of the average quality of water at a given location over a certain time period. The type of sample selected for analysis depends on the type of water that will be sampled, the tests that will be run, and the use to which the test results will be put.

Grab samples

A GRAB SAMPLE is a single volume of water collected all at one time from a single place. For example, the operator sampling raw water might fill a sample bottle by dipping it into a creek or tank of water. Or, to sample water in the distribution system, he might open a faucet or sample tap and fill the bottle. In either case, he would have taken a grab sample.

The quality of a grab sample represents the quality of the water only at the time the sample was taken. If a grab sample is taken from a water source where quality is slow to change (such as ground water), then the sample is usually considered representative of the water quality over a long period of time, not just the instant the sample was taken. On the other hand, if a grab sample is taken from a source where some quality changes occur hourly, daily, weekly, or seasonally (for example, river water), then the time period over which the sample can be considered representative depends on the constituents being measured. For example, if dissolved oxygen (DO) or pH levels in the river are being measured, then the sample is only representative of the water at the instant being sampled, because DO and pH can change hourly. However, if total dissolved solids (TDS) or an inorganic element such as chloride or fluoride is being measured, then the sample may be considered representative of the water quality that week—probably even that month—since these constituents tend to change slowly.

The decision as to how representative a grab sample will be for a given constituent must be based on the history of water quality for the water being sampled, as shown by the records of previous test results over a period of several years. These records will help in determining how often grab samples should be collected to adequately represent the water quality and to monitor the changes in that quality.

Figures 2-1 and 2-2 illustrate this point. Figure 2-1, which shows how DO in surface water might change throughout a 24-hour day, indicates that a grab sample would only be representative of DO levels at the time the sample was taken. Yet Figure 2-2 shows that for tests of TDS levels in the same water, a grab

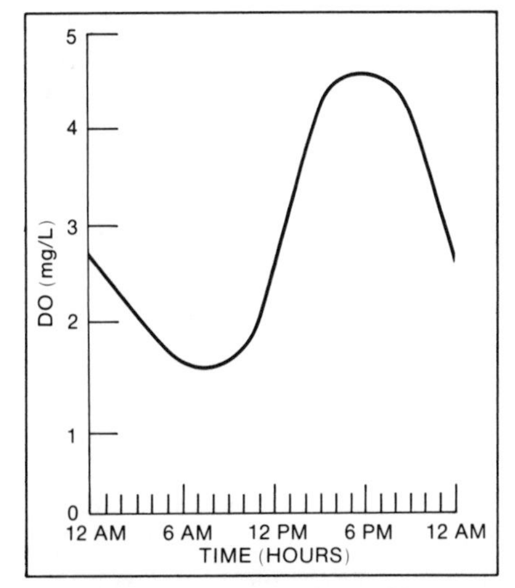

Figure 2-1. Example: Hourly Changes in DO for One Surface-Water Source

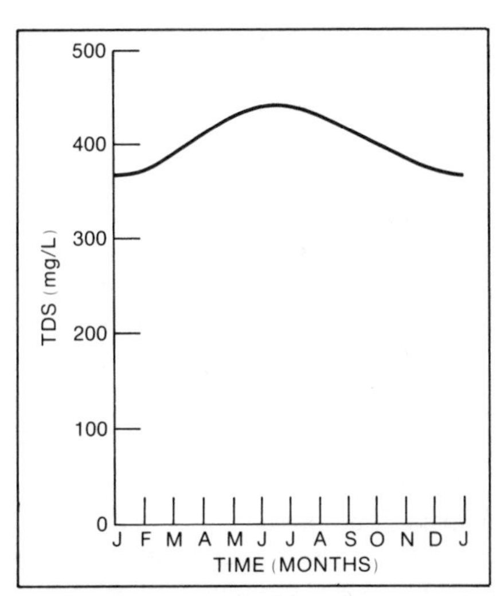

Figure 2-2. Example: Monthly Changes in TDS for One Surface-Water Source

sample collected one afternoon in April could be assumed to adequately represent water quality not only for that day but perhaps for the entire month, because TDS levels in the water have been very slow to change.

Composite Samples

A COMPOSITE SAMPLE is a series of grab samples taken at different times from the same sampling point and mixed together. A composite sample can be made up of several equal-volume grab samples, taken at different times—this is called a TIME COMPOSITE. Or, a composite sample can consist of a mixture of grab samples of different volumes taken at different times, with the volume of each grab sample varying, depending on the flow rate at the time of sampling—this is called a FLOW-PROPORTIONAL COMPOSITE.

Composite samples are used to determine an average concentration. For example, to determine the average level of fluoride (F) leaving the treatment plant on a given day, an operator could collect and test 24 separate grab samples and average the 24 results. Or, the operator could collect 24 samples, combine them, and test the one composite sample. Although time spent in sampling would be the same in both cases, considerable laboratory time would be saved by using the composite sample. One example of the usefulness of composite sampling is in the sampling of radiochemicals for monitoring under the USEPA drinking water regulations (the NIPDWRs, described in detail in Module 1, Drinking Water Standards). The regulations specify that systems may either analyze an annual composite of four consecutive quarterly samples or average the results of four grab samples obtained at quarterly intervals. Using the composite sampling alternative will reduce the cost of testing to one quarter of what it would be if grab samples were tested individually.

Not all constituents are stable enough to be stored during the period necessary to build up a composite sample. Bacteriological samples should never be composited. The number of coliform bacteria in the samples begins to change immediately after sample collection; also, the sterile conditions required in the coliform sample bottles could not be maintained with composite sampling procedures. Therefore, bacteriological testing should begin as soon as possible after the sample is taken—within 30 hours as a maximum. Temperature, pH, chlorine residual, and dissolved gases (including DO) are other characteristics that usually require grab sampling and immediate testing because their levels of concentration can change rapidly during storage.

2-2. Sample Volumes

The volume of a sample depends on what tests it will be used for. Table B.1 in Appendix B summarizes recommended sample volumes for all the water-quality tests covered in this text, including all those required under the NIPDWRs. However, since sample-volume requirements can vary depending on the testing procedures used, it is always advisable to check with the laboratory to determine the exact sample volume it requires.

It is not necessary to collect a separate sample for each chemical constituent that must be monitored. If the recommended holding period, type of container, and preservation methods are the same for several chemicals, then the levels of those constituents can be determined from a single sample of sufficient volume. This simplifies the collection procedure a great deal. For example, only one sample needs to be collected for analyses of all six organic chemicals that must be monitored under the NIPDWRs.

2-3. Sample-Point Selection

The selection of representative sample points is one of the important steps in developing a sampling procedure that will accurately reflect water quality. The criteria used in selecting sample points will depend on the type of water being sampled and the reason for sampling. The following section reviews sample-point selection criteria for sampling in three general areas:

- Raw-water supply
- Treatment plant
- Distribution system.

Selecting Raw-Water Sample Points

The collection points for raw-water samples depend on the type of raw-water system being sampled. There are at least four general types: (1) raw-water transmission lines; (2) ground water (wells); (3) streams or rivers (water courses); and (4) lakes or reservoirs (water bodies).

Raw-water transmission lines and ground-water sources are generally sampled directly from the transmission or well-discharge line. After a sampling

Figure 2-3. Sample Cock Attached to Pipeline for Sampling

point is selected, the transmission or well-discharge line is equipped with a sample cock—a small valve attached to the pipeline (Figure 2-3). The valves must be fully opened before sampling to flush out any standing water and accumulated sediment.

Selecting a representative sample point in a stream may be difficult. The sample must be taken far enough away from the bank to avoid eddys and stagnant or very slowly moving waters; to avoid collecting sediment or floating debris in the sample, a relatively deep point should be selected. It may be that the mainstream of flow can only be sampled by wading or from a boat. If wading, the sample container should always be held upstream of the person taking the sample to avoid collecting sediment.

Lakes and reservoirs present a challenging sample-point selection task. If the purpose of the sampling program is to determine the quality of the water leaving the reservoir, a sample point at the discharge from the water body should be selected. However, if the quality of water in the water body itself is to be tested, then the choice of a sampling point is more difficult. Figures 2-4 and 2-5 show suggested sample-point locations in a natural lake and a reservoir for both routine and TRANSECT sampling. Transect sampling is usually performed as part of a detailed study of the physical, chemical, and biological characteristics of a water body.

Selecting Sample Points Within the Treatment Plant

Selection of in-plant sample points is an important step in developing an overall process-control program for a water treatment plant. Samples from the points selected can be tested to determine the efficiency of the various treatment processes, and the test results will help to indicate operational changes that can increase removal efficiencies or reduce operating costs. Treatment plants vary

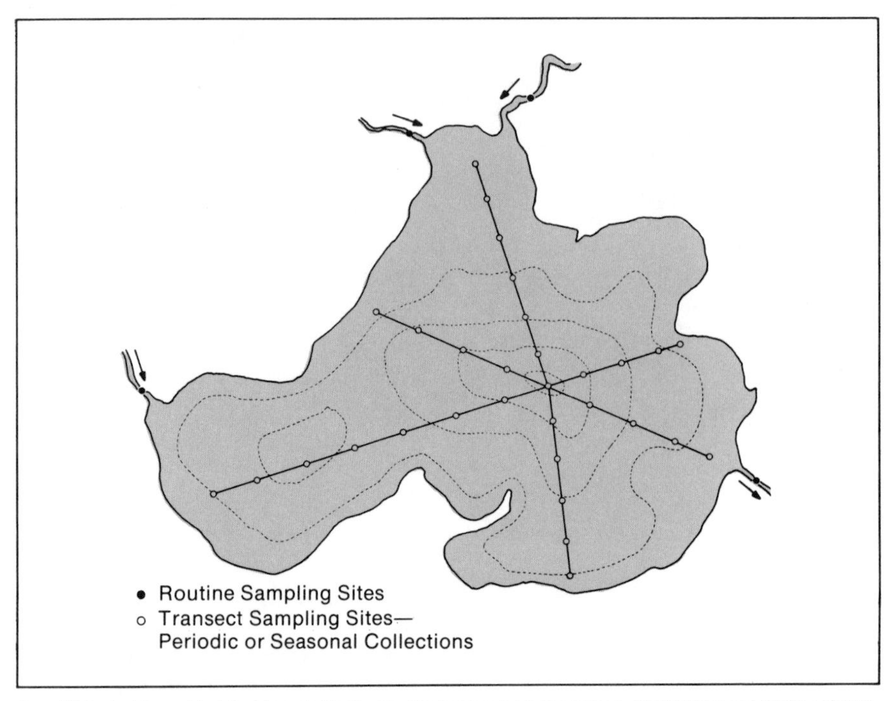

From *Biological Associated Problems in Freshwater Environments* by Kenneth M. Mackenthun and William Marcus Ingram. US Dept. of the Interior, Federal Water Pollution Control Admin.

Figure 2-4. Routine and Transect Sample Points in a Natural Lake

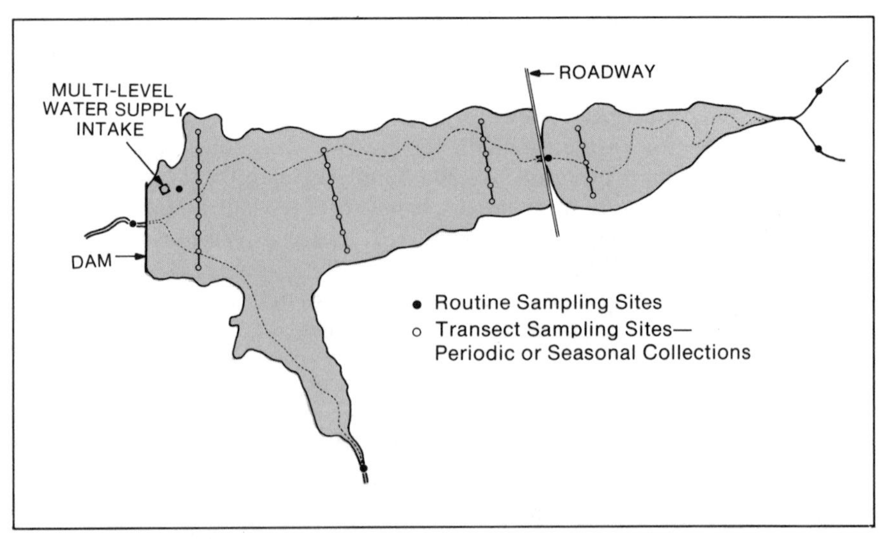

From *Biological Associated Problems in Freshwater Environments* by Kenneth M. Mackenthun and William Marcus Ingram. US Dept. of the Interior, Federal Water Pollution Control Admin.

Figure 2-5. Routine and Transect Sample Points in a Reservoir

widely as to the kinds of treatment processes used and the arrangement of the processes. Consequently, it is not possible in this text to identify precise sample-point locations for all cases. However, using Figure 2-6, general sample-point locations can be identified.

In-plant sample points should be established at every point where, because of a treatment method or group of methods, a measurable change is expected in the treated-water quality. For example, between sample points 1 and 2 of Figure 2-6, test results should show a reduction in algae and the associated taste and odors (the result of chemical pretreatment), a reduction in sediment load (the result of presedimentation), and a reduction in debris (the result of screening). Between points 2 and 3, aeration should cause a significant reduction in dissolved gases and some reduction in iron and manganese. Between points 3 and 4, the combined effects of coagulation, flocculation, and sedimentation should cause a reduction in turbidity and color.

Water-quality changes between sample points 4 and 5 will allow the operator to monitor the effectiveness of the softening process. Sample points 5 and 6 allow monitoring of the efficiency of filtration in removing turbidity and previously oxidized iron and manganese (by aeration), as well as a reduction in bacteria (measured by the standard plate count). Sampling at points 6 and 7 will indicate the efficiency of the adsorption process in removing organic chemicals.

As water reaches the final stages of the treatment process, single sample points may be used to check the performance of a process or monitor compliance with an MCL. For example, point 8 is used for the measurement of fluoride concentration, point 9 for final pH and alkalinity, and point 10 for coliform bacteria, chlorine residual, and turbidity in the finished water.

When selecting in-plant sample collection points, certain precautions should be kept in mind. Points immediately downstream from chemical additions should be avoided since proper mixing and reaction may not have taken place. Samples should always be taken from the main stream of flow, avoiding areas of standing water, algae mats, or other floating debris.

Finished-water sample points are normally established downstream of the final treatment process, at or just before the point where the water enters the distribution system, such as the discharge point from the clear well. For example, turbidity samples required by the NIPDWRs must be collected before the water enters the distribution system. This will prevent high turbidity readings due to scale or sediment in the distribution piping.

Selecting Distribution-System Sample Points

Distribution-system sample points are used to determine the quality of the water delivered to the consumers. Samples may be of significantly different quality than samples of finished water taken at the plant. Corrosion in the distribution-system pipelines can cause increases in water color, turbidity, and taste-and-odor problems. More seriously, a cross connection between the distribution system and a source of contamination can result in chemical poisoning or outbreaks of waterborne disease.

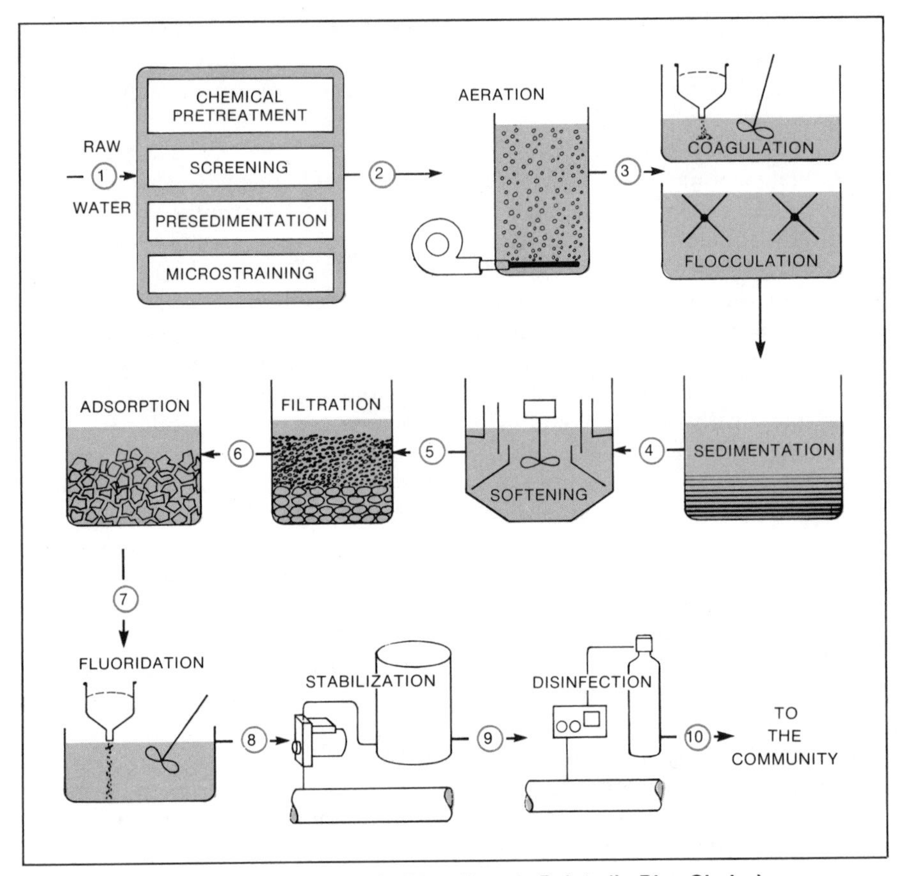

Figure 2-6. Suggested In-Plant Sample Points (In Blue Circles)

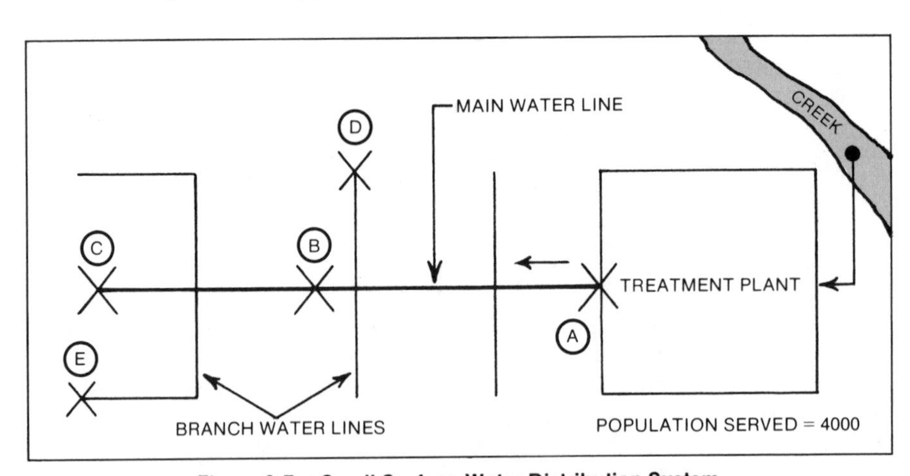

Figure 2-7. Small Surface-Water Distribution System

Most of the samples collected from the distribution system will be used to test for coliform bacteria and chlorine residual. Others may be used to determine quality changes, and still others will be used to test for compliance with the inorganic, organic, and radiochemical MCLs required under the applicable drinking water standards.

Distribution-system sampling should always be performed at points representative of conditions within the system. The number and location of sample points should be established to ensure compliance with the applicable coliform-testing requirements of the state or federal drinking water regulations.

To comply with the NIPDWRs, every surface-water system with a population of 25 to 1000 must have at least two sampling points: one for turbidity testing at the point where the water enters the distribution system, and one for coliform testing at a consumer's faucet representative of conditions within the system. Systems drawing surface water from more than one source will require additional turbidity sampling points, and systems serving populations greater than 1000 will require more than one coliform sampling point. These sample points can also serve for other distribution system sampling.

The two major considerations in determining the number and location of sampling points are that they should be:

- Representative of each different source of water entering the system
- Representative of conditions within the system, such as dead ends, loops, storage facilities, and pressure zones.

The precise location of sampling points will depend on the configuration of the distribution system. The following example should provide some overall guidance on sample-point selection for the distribution system.

Example 1. Figure 2-7 represents a small surface-water distribution system serving a population of 4000. It is a typical small branch system having one main water line and several branch or dead-end water lines. For this system, a single sampling point *A* is sufficient for turbidity sampling. This point is representative of all treated water entering the distribution system.

The NIPDWRs require a minimum of four bacteriological samples per month for a community of 4000 (as discussed in Module 1, Drinking Water Standards). To be representative of the system, the samples should be taken at four different points: *B, C, D,* and *E*. Point *B* is representative of water in the main line, whereas point *C* represents water quality in the main-line dead end. Points *D* and *E* were selected to produce samples representative of a branch line and a branch-line dead end, respectively.

Consideration of how often and at what times these points are sampled is also necessary to ensure the samples are representative of conditions within the distribution system. Although the minimum requirement of four samples per month could be met by collecting samples from all points on one day, this sampling frequency would not produce samples that represented bacteriological conditions within the system throughout the month. A better program would be to sample points *B* and *E* at mid-month and points *C* and *D* at the end of the

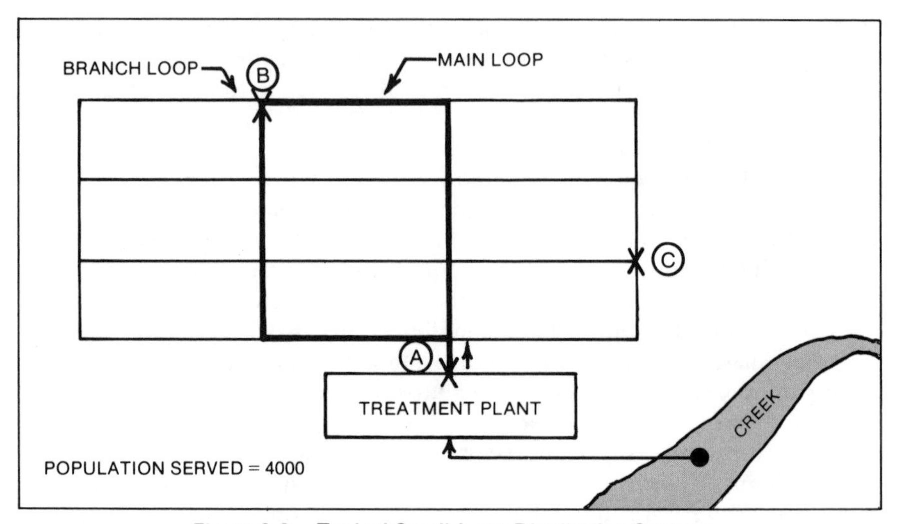

Figure 2-8. Typical Small-Loop Distribution System

month. Representative sampling must be both representative in location and representative in time.

Although this type of program is adequate to meet minimum monitoring requirements of the NIPDWRs, good operating practices would include sampling at each dead end at perhaps monthly intervals and probably several additional sampling points within the distribution system so that samples would be taken each week. The exact number and location of these operational sampling points will depend on the characteristics of a system.

Chlorine residual samples should be taken from each sample point at the time bacteriological samples are collected. Sampling for routine water chemistry, along with the required inorganics, organics, and radiological sampling, also can be conducted at one of the coliform sampling points (the consumer's faucet). Sampling for a similar system using a ground-water source would be the same, except that turbidity sampling is generally not required.

Example 2. Figure 2-8 illustrates a typical small-loop distribution system having one main loop and several branch loops. This community water system serves surface water to a population of 4000, similar to the system in Figure 2-7, and the required samples and sampling frequency are the same.

For the system shown in Figure 2-8, one turbidity sample point is sufficient (point *A*) because that point is representative of all treated water entering the distribution system. It must be sampled daily. For bacteriological sampling, two sampling points are adequate: points *B* and *C*. Point *B* is representative of water in the main-line loop, and point *C* is representative of water in one of the branch-line loops. To produce the required minimum of four samples per month, points *B* and *C* can be sampled on alternate weeks (one sample each week). This program will produce samples that are representative in time and

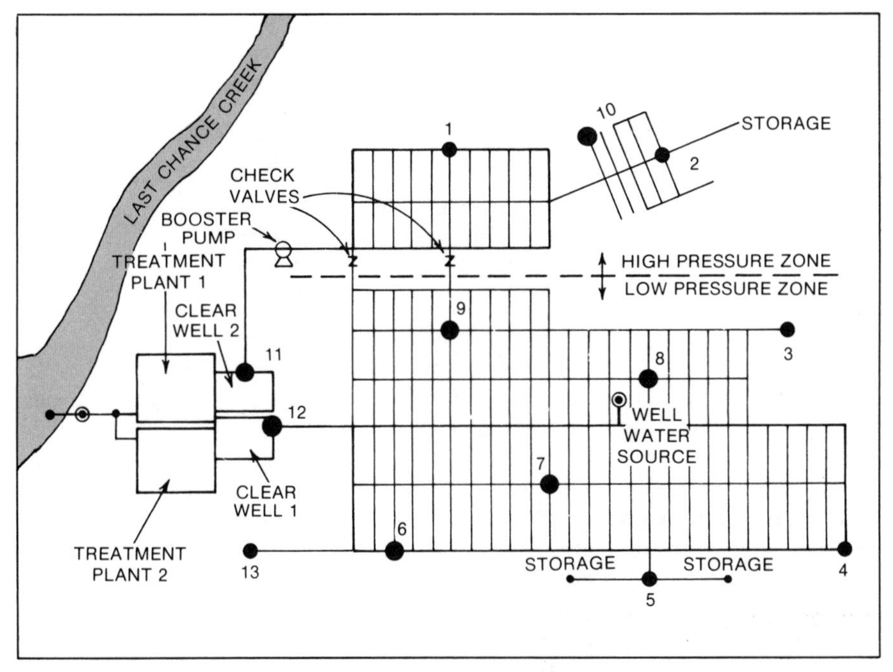

Figure 2-9. Sample City, Mont. Distribution System

location. However, good operating practice would include two to three times this number of samples, depending on the characteristics of the particular system.

As with the system in the previous example, chlorine residual samples should be taken whenever bacteriological sampling is performed, and other routine samples also can be taken from the coliform sampling points.

Example 3. As a final example of sample-point selection, consider the Sample City, Mont., water system shown in Figure 2-9. The system, serving a population of 17,440, obtains its water from Last Chance Creek and one well. The distribution system has the features of both the branch and loop systems shown in Figures 2-7 and 2-8.

To determine sample-point locations, the following four questions should be considered:

1. What tests must be run?
2. From what locations will the samples be collected?
3. How often must the samples be taken?
4. How many sampling points will be needed?

The answers to questions 1 and 3—what tests must be run and how often—will vary from state to state. For Sample City, the testing requirements specified as a minimum by the NIPDWRs are as follows:

Tests	*Sampling Location*	*Sampling Frequency*
Inorganics	Consumer's faucet	Once/year
Organics	Consumer's faucet	Once/3 years
Turbidity	At point(s) where water enters the distribution system	Daily
Coliform bacteria	Consumer's faucet	20/month
Radiochemicals	Consumer's faucet	4 quarterly samples every 4 years.

Additional testing may be required by state regulations or may be necessary for the utility's own quality control program. Other tests that might be run include chlorine residual, odor, color, pH, TDS, iron, manganese, and standard plate count.

Once the tests and test frequency are determined, the number and specific location of sampling points must be selected. The NIPDWRs require a turbidity sample to be taken at each point where surface water enters the distribution system. Since waters from parallel treatment plants enter two separate clear wells in Figure 2-9, two turbidity sampling points are required (points 11 and 12). Notice there is no sampling point shown where the well-water source enters the system; this is because ground-water sources need not be monitored for turbidity.

In selecting sample points that will be representative for coliform analysis, a variety of factors must be considered:

- Uniform distribution of the sample points through the system
- Location of sample points in both loops and branches
- Adequate representation of sample points within each pressure zone
- Location of points so that water coming from storage tanks can be sampled
- For systems having more than one water source, location of sample points in relative proportion to the number of people served by each source.

Using these basic considerations, bacteriological sample points can be selected. A community the size of Sample City must have 20 coliform bacteria samples tested per month, according to the NIPDWRs. After carefully reviewing the configuration of Sample City's distribution system, 10 coliform bacteria sample sites are selected. The reason for selecting each point is noted in the following paragraphs.

1. Point 1 is on the main loop in the high-pressure zone; it should produce representative samples for that part of the system.
2. Point 2 is on the branch loop in the high-pressure zone, representative of storage flow to the system.

3. Point 3 is on a dead end. Many authorities advise against dead-end sampling points because they do not produce representative samples. However, consumers do take water from branch-line dead ends. In the case of Sample City, there are seven branch-line dead ends, no doubt serving significant numbers of consumers. It is representative to have one or two sample points on these branch lines at or near the end.
   If there are indications of bacteriological problems in sampling branch-line dead ends, then hydrants and blow-off valves should be flushed and branch lines resampled immediately to determine if the problem has been corrected. If the problem persists, additional investigation is needed to locate the source of contamination.
4. Point 4 is located on the main loop of the low-pressure zone and represents water from treatment plant 2, the well, the storage tanks, or any combination (depending on system demand at sampling time).
5. Point 5 samples the quality of water flowing into the system from storage.

6-9. Points 6 through 9 were selected by uniformly distributing points in the low-pressure zone, the zone that serves the major part of the community.

10. Point 10 was selected as representative of a branch-line dead end in the high-pressure zone, just as point 3 was selected in the low-pressure zone.

11-12. (Points 11 and 12, as stated previously, are used as turbidity monitoring points.)

13. Point 13 was added to monitor a dead-end branch that is fairly isolated from other sampling points, yet serves a large population.

Sample faucets. Once a representative sample point has been located on the distribution system map, a specific sample faucet must be selected. These faucets can be located:

- Inside a public building, such as a fire station or school building
- Inside the home of the operator
- Inside the homes of other consumers.

Faucets selected should be on lines connected directly to the main. Only cold-water faucets should be used for sample collection. The sampling faucet must not be located too closely to a sink bottom—contaminated water or soil may be present on such faucet exteriors, and it is difficult to place a collection bottle beneath them without touching the neck interior against the outside faucet surface. Samples should not be taken from:

- Swivel faucets
- Leaking faucets, which allow water to flow out from around the stem of the valve and down the outside of the faucet
- Lines with home water-treatment units, including water softeners
- Drinking fountains.

Aerators, strainers, and hose attachments on nonswivel water faucets can harbor a significant bacterial population and should always be removed before sampling.

Once a representative sample point has been selected, it should be marked or described so that it can be easily located for future sample collection.

2-4. Sample Collection

This section explains the steps to follow when collecting:

- Raw-water samples
- In-plant samples
- Distribution-system samples
- Special-purpose samples.

The steps outlined represent general sample-collection procedures that should be followed regardless of the constituent to be tested. Special sample-collection procedures required for certain tests are described in Module 4, Microbiological Tests, and Module 5, Physical/Chemical Tests.

A clean sample bottle fitted with a lid should always be used. With few exceptions, the bottle can be either glass or plastic. During sample collection, the bottle lid should be removed and held threads down. The lid can be easily contaminated if the inside is touched or if the lid is set face down or placed in a pocket. A contaminated lid can contaminate the sample.

Raw-Water Sample Collection

A clean, wide-mouth sampling bottle should be used. The bottle should not be rinsed—this is especially important if the bottle has been pretreated, sterilized, or contains a preservative. The open bottle should be held near its base and plunged neck downward below the surface. The bottle should then be turned until the neck points slightly upward with the mouth directed toward the current. Care must be taken to avoid floating debris and sediment. When sampling in a water body with no current, the bottle can be scooped forward to create a current. Once the bottle is filled, it should be brought to the surface, capped, and labeled.

When wading to take a sample, the sample bottle should be submerged upstream from the person taking the sample. If a boat is used for stream sampling, the sample should be taken on the upstream side.

When sampling from a large boat or a bridge, the sample bottle should be placed in a weighted frame that holds the bottle securely. The opened bottle and holder are then slowly lowered toward the water with a small diameter rope. When the bottle approaches the water, the unit is dropped quickly into the water. Slack should not be allowed in the rope since the bottle could hit the bottom and break or pick up mud and silt. After the bottle is filled, it is pulled in and the bottle is capped and labeled.

In-Plant Sample Collection

When sampling from an open tank or basin or in an open channel of moving water, the sample-collection procedure is essentially the same as described previously for raw-water sampling.

Many plants equipped with laboratories have sample taps installed in the laboratory. These faucets provide a continuous flow of water from various locations in the treatment plant. To collect a sample, the operator or laboratory technician draws the required volume from the sample tap. Figure 2-10 shows a typical bank of sample faucets in a laboratory.

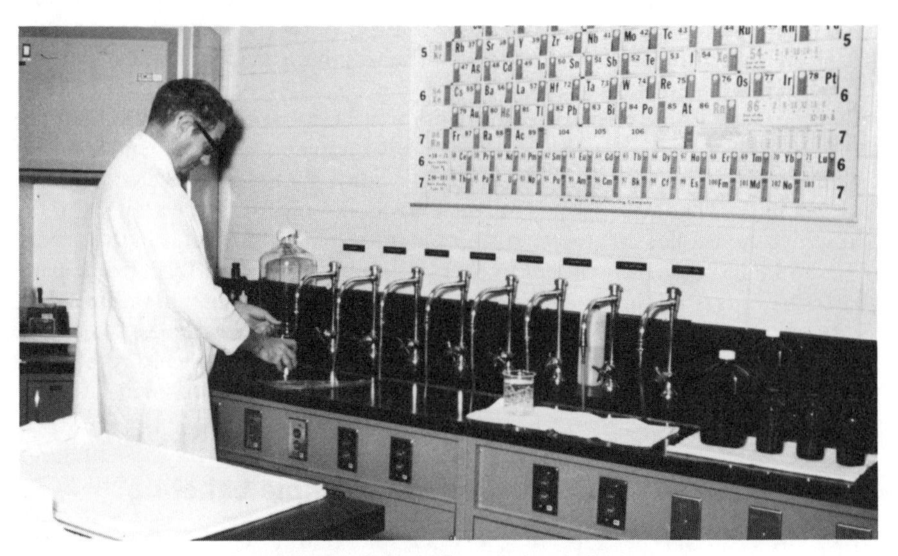

Figure 2-10. Sample Faucets in a Laboratory

Distribution-System Sample Collection

Once the sample locations have been selected, sample collection consists of a few, simple, carefully performed steps.

First, the faucet is turned on and set to produce a steady, moderate flow of water. If a steady flow of water cannot be obtained, the tap should not be used. The faucet is allowed to run long enough to flush any stagnant water from the service line. This usually takes 2 to 5 min. The line is usually clear when the water temperature drops. Without changing the flow, the sample is then collected. The sample-bottle lid should be held threads down during sample collection and replaced on the bottle immediately. The final step is to label the bottle.

FLAMING the outside of a water faucet is not recommended. The flame from a match or an alcohol-soaked swab is not hot enough and does not last long enough to kill all the bacteria on the outside of the faucet, and a more intense heat, as from a propane torch, might damage the valve washer and seat or could set nearby combustible material on fire.

Special-Purpose Samples

Occasionally, a water utility may have need to collect samples to be used for special testing purposes. Sampling procedures in such cases depend on the reason for the sample.

For example, perhaps a consumer has complained about a taste, odor, or color in the water. In such a case, samples should be taken from the consumer's faucet to determine the source of the problem. The faucet is opened and a sample taken immediately. This sample represents the quality of water standing in the service line. The faucet is then allowed to run for 2 to 5 min so that the standing water in the service line is completely flushed out and a second sample is taken. The second sample is fresh from the distribution system. Comparing test results from the two samples will help identify the origin of the problem causing the consumer complaint.

Often taste, odor, or color complaints will be caused by conditions in the consumer's hot-water heater, water softener, or home water-treatment device. If the hot-water supply is suspected, the first sample should be collected from the hot-water tap. The tap is turned on and allowed to run until the water is hot; the sample is then collected. A second sample representing the water in the service line should be taken from the cold-water tap as previously described. Comparing the test results from the two samples will help identify the origin of the problem.

There are many other reasons for taking special-purpose samples. The above examples point out the importance of knowing what the sample test results will be used for, so that a sample can be collected that will be representative of the conditions tested.

2-5. Record Keeping and Sample Labeling

Records should be kept of every sample that is collected. A sample-identification label or tag should be made out at the time of collection. Each label or tag should include at least the following information:

- Date sampled
- Time sampled
- Location sampled
- Type of sample—grab or composite
- Tests to be run
- Name of person sampling
- Preservatives used
- Bottle number.

The laboratory should always be provided with a sample that is clearly labeled. A sample label appears in Figure 2-11. The information on the label should also be entered on a record-keeping form that is maintained as a permanent part of the water system's records.

CITY OR FIRM Ashland COLLECTED BY VLS

DATE & TIME SAMPLED 11:30 AM 1-12-78

LOCATION House, 1208 Main, Kitchen

TESTS Total Coliform

CHEMICAL ADDITION Sodium Thiosulfate

SAMPLE TYPE Distribution

BOTTLE NO. 118

Figure 2-11. Sample Label

2-6. Sample Preservation, Transportation, and Storage

Samples cannot always be tested immediately after they are sampled. To ensure that the levels of the constituent remain unchanged until testing is performed requires careful attention to techniques of sample preservation, transportation, and storage.

Preservation

Once a sample has been collected, its quality begins to change, because of chemical and biological activity in the water. Some characteristics, such as alkalinity, pH, dissolved gases, nitrogen, and odor, can change quickly and quite extensively. Other characteristics, such as pesticides and radiochemicals, change more slowly and much less noticeably.

If these changes are allowed to occur at their normal rate, the test results will not represent the quality of the water that was sampled. To solve this problem, sample-preservation techniques have been developed that slow the chemical and biological activity in the sample, allowing it to be transported to the laboratory and tested before significant changes occur.

Sample preservation usually involves two basic steps:

- Refrigeration
- pH adjustment.

Table B.1 in Appendix B shows the recommended preservation steps to be taken for each water-quality test covered in this text.

Transportation

If samples are to be analyzed by a state or private laboratory in a distant city, a method of shipment must be found that will ensure that the samples arrive at the lab prior to the expiration of the allowed storage time. Usually the US mail or a commerical package-shipping service is the best way to ship samples. When shipping, make sure that the bottle caps are tight to prevent leakage and that the samples are packed in a sturdy container with enough cushioning material to prevent breakage.

Sample Storage

It is not always possible to analyze a sample immediately upon its arrival at the laboratory. Ongoing tests may have to be completed or test equipment set up before a new test can be begun. All of these situations require that the samples be stored to await testing.

The amount of time a sample can be stored depends on the constituent's stability and on whether the constituent can be preserved—that is, whether a chemical can be added to slow down or stop changes. Not all constituents can be preserved. For example, samples to be tested for DO, temperature, turbidity, and chlorine residual are very unstable—they cannot be effectively preserved and must be analyzed immediately after sampling. Samples for those characteristics, therefore, must not be stored. On the other hand, some constituents are very stable or they can be effectively preserved, thereby allowing for some storage time. Table B.1 in Appendix B summarizes the recommended storage times for some of the more common water-quality tests. Storage times, when allowed, vary from as little as 6 hours to an indefinite period, depending on the test.

Selected Supplementary Readings

Handbook for Sampling and Sample Preservation of Water and Wastewater. (EPA-600/4-76-049) USEPA, Washington, D.C. (1976).
Available through National Technical Information Service (NTIS) Springfield, VA 22161.

Methods for Collection and Analysis of Water Samples for Dissolved Minerals and Gases. US Geological Survey, US Government Printing Office, Washington, D.C. (1974).

Safe Water—A Fact Book on the Safe Drinking Water Act for Non-Community Water Systems. AWWA. Denver, Colo. (1979).

Smith, R.A. *Proper Bacteriological Sampling of a Domestic Water Supply Distribution System.* Water Technology Conf. Proc., Paper No. 3B-4. AWWA. Denver, Colo. (Dec. 1977).

Safe Drinking Water Act Handbook for Water System Operators. AWWA. Denver, Colo. (1978).

Safe Drinking Water Act Self-Study Handbook for Community Water Systems. AWWA. Denver, Colo. (1978).

Glossary Terms Introduced in Module 2

(Terms are defined in the Glossary at the back of the book.)

Composite sample
Flaming
Flow-proportional composite
Grab sample
Representative sample
Time composite
Transect

Review Questions

(Answers to Review Questions are given at the back of the book.)

1. What are the two general types of samples?

2. What is the difference between a grab sample and a composite sample?

3. Aside from the constituents for which a sample is to be tested, what may affect the volume of the sample?

4. Using Table B.1 in Appendix B, give appropriate volumes for the following tests: hardness, pH, total dissolved solids, chlorine, fluoride.

5. What are the four general types of raw-water sample points?

6. What precautions should be taken in selecting in-plant sample points?

7. List two constituents for which samples from the distribution system are regularly tested.

8. What test is commonly taken at the same time as bacteriological samples?

9. What problems most commonly occur in dead ends?

10. In distribution system sampling, is the sample taken from the faucet immediately? Why or why not?

11. What samples should be taken in testing for taste or odor from a hot-water heater?

12. What eight pieces of information should appear on every sample-identification label?

13. List two characteristics of water that change slowly.

14. What are two methods used to preserve samples?

15. What two things determine the length of time a sample can be stored?

Study Problems and Exercises

1. Your water system has just developed a new surface-water source and a water treatment plant has been constructed. The new water source is a stream, which originates in a heavily wooded watershed containing abandoned gold mines. Treatment processes consist of coagulation, flocculation, sedimentation, filtration, fluoridation, and chlorination.

   The city manager has asked you to develop a sampling program that will ensure effective plant operations.

   (a) What constituents would you sample for in the source water?

   (b) At what frequency would you take the samples? Specify if different frequencies would be used for different constituents.

   (c) What locations would you select to take the raw-water samples?

   (d) What constituents would you sample for in the treatment plant?

   (e) At what frequency would you take the samples? Specify if different frequencies would be used for different constituents.

   (f) Sketch where your sampling points would be located in the treatment plant and list the constituents that would be sampled at each sampling point.

Module 3

Water Laboratory Equipment and Instruments

Water plant operators are accustomed to large, rugged equipment used for measuring and handling tons of material or millions of gallons of water. In a water treatment plant laboratory, fragile laboratory equipment and delicate instruments are used to detect and precisely measure extremely small quantities of contaminants. This module discusses the kinds of laboratory equipment and instruments needed to conduct most of the tests discussed in Modules 4 and 5. The sections covered are (1) labware (common glass and plastic utensils), (2) major laboratory equipment (mechanical equipment of prime importance to a test or procedure), (3) support laboratory equipment (mechanical equipment of secondary importance to a test or procedure), and (4) laboratory instruments (certain physical or electronic devices used in specific water quality test procedures).

After completing this module you should be able to

- Recognize general labware and laboratory equipment.
- Recognize analytical instruments commonly used in laboratory work.
- Recommend basic labware and instruments necessary to conduct routine process control tests.

The proper use and operation of laboratory equipment and instruments necessitates the development of proper laboratory techniques. These techniques can vary between pieces of equipment that perform the same analyses but are manufactured by different companies. For this reason, no in-depth discussion of laboratory techniques will be included in this module. A good way to learn proper lab techniques is to attend training sessions offered by state health or environmental agencies, large water utilities, operators' associations, or local community colleges.

A prime concern in any laboratory is safety. State occupational safety and health agencies can provide regulations and valuable advice on laboratory safety. This information is particularly important when setting up a lab.

3-1. Labware

The most suitable material for labware (bottles, beakers, etc.) is heat-resistant BOROSILICATE GLASS, commonly sold under the trade names "Pyrex" or "Kimax."

Heat-resistant glass can be repeatedly AUTOCLAVED (sterilized at elevated temperature and pressure), can be heated over open flames without shattering, and can also withstand heat generated from chemical reactions. However, rapid heating and cooling can weaken even heat-resistant glass, eventually causing it to crack or shatter.

Plastic is the second most common labware material. It is suitable for many laboratory purposes. Some types of plastic are resistant to high temperatures and can be autoclaved.

The extensive use of plastic labware is a matter of choice. Plastic labware is unbreakable and, in some cases, disposable, which eliminates the need for laborious cleaning procedures. However, reusable plastic is harder to clean and cannot be used for all chemical analyses. For example, plastic labware should not be used in preparing samples for pesticide analysis because the plastic will adsorb the pesticide, causing erroneous results. In certain other tests, such as extractions using organic solvents, the chemicals used will deteriorate plastic almost immediately. Additionally, plastic labware is easily scratched and marred, and it becomes more cloudy with use than glass labware. Therefore, although plastic labware can be used extensively in the laboratory, it cannot and should not replace glass labware entirely.

Another suitable labware material is soft (non-heat-resistant) glass, which can be used to store some dry chemical reagents, such as calcium, magnesium, sulfate, and chloride. This material is not recommended for extensive laboratory use, since it breaks easily and cannot be used with heat.

Some of the common types of laboratory containers are described in the following paragraphs.

Beakers. BEAKERS are glass jars with open tops, and vertical side walls, and pouring lips that simplify pouring of liquids (Figure 3-1). Common laboratory beakers range in size from 25 to 2000 mL. The 250- and 500-mL sizes are the most popular. Beakers are used as mixing vessels for most chemical analyses. Consequently, an ample supply of various sizes should always be kept on hand.

Burets. BURETS are glass tubes (Figure 3-2), graduated over part of their length and fitted with a stopcock. The most common sizes are 10, 25, and 50 mL. The graduations are normally in tenths of a millilitre. Burets are designed for dispensing solutions during TITRATION, a procedure commonly used when determining the concentration of a substance in solution. Certain manufacturers

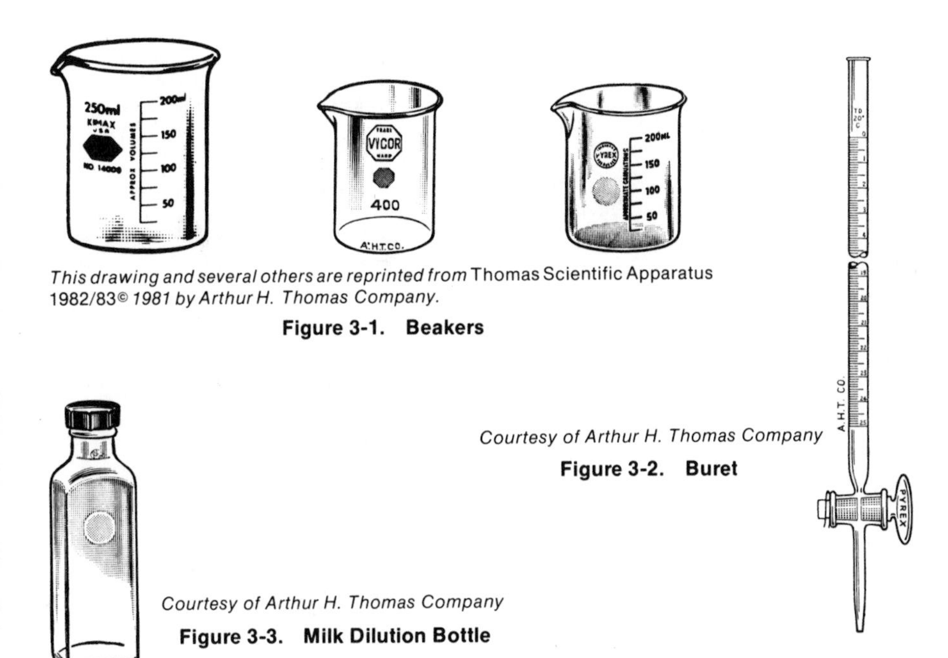

This drawing and several others are reprinted from Thomas Scientific Apparatus 1982/83© *1981 by Arthur H. Thomas Company.*

Figure 3-1. Beakers

Courtesy of Arthur H. Thomas Company

Figure 3-2. Buret

Courtesy of Arthur H. Thomas Company

Figure 3-3. Milk Dilution Bottle

now produce plastic burets, which are much more durable than glass burets. Plastic burets are especially useful for field tests.

Dilution bottles. DILUTION BOTTLES, also known as MILK DILUTION BOTTLES or FRENCH SQUARES, are heat-resistant glass bottles used for diluting bacteriological samples for analysis. The bottles are square (Figure 3-3) with narrow mouths threaded to receive a screw cap. All bottles have a 160-mL capacity with a mark at the 99-mL level to facilitate 1-to-100-mL dilutions of a sample.

Flasks. There are many types of FLASKS (Figure 3-4), each with its own specific name and use. Some names specify their use—"boiling," "distilling," "filter." Other names specify the test they are used for, such as "Kjeldahl." All flasks have narrow necks. Erlenmeyer and volumetric flasks are the most common types.

ERLENMEYER FLASKS are one of the most common pieces of labware. Characterized by their bell shape, they are recommended for mixing or heating chemicals since they minimize splashing. They are also frequently used for preparing and storing culture media. They range in size from 100 to 4000 mL.

VOLUMETRIC FLASKS have long, narrow necks. They range in size from 10 to 2000 mL; the level at which the flask's capacity is reached is indicated by an etched ring around the neck. Volumetric flasks are used for preparing and diluting standard solutions. Since these flasks are designed for measuring, they should not be used for long-term storage of solutions.

The other types of flasks shown in Figure 3-4 have various specialized uses that will not be discussed in detail here.

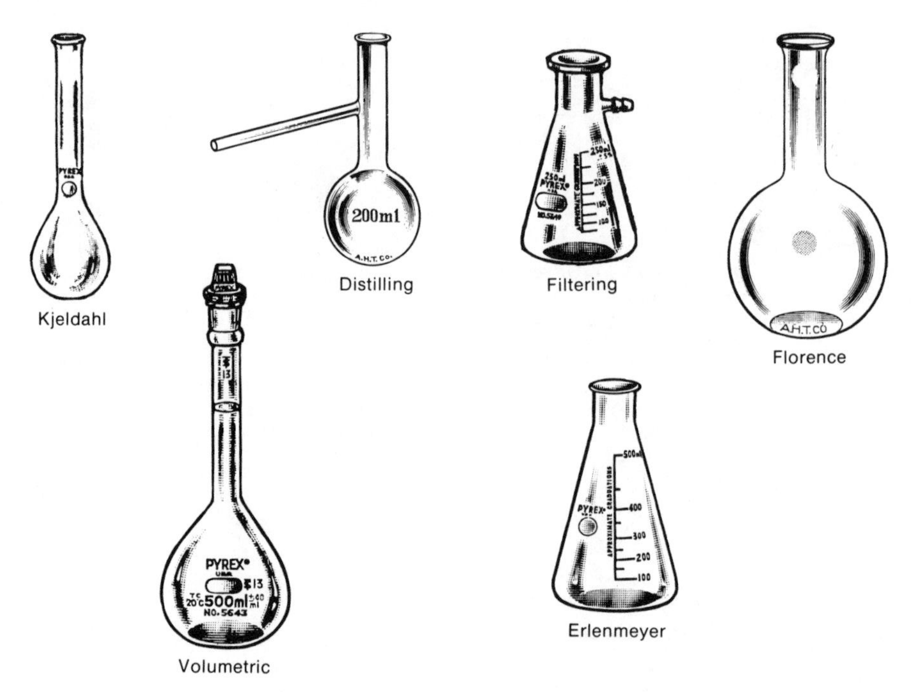

Courtesy of Arthur H. Thomas Company

Figure 3-4. Flasks

Funnels. FUNNELS are a common piece of laboratory equipment. Four of the most frequently used funnels are shown in Figure 3-5. The most common type, the general-purpose funnel, is used to transfer liquids into bottles or to hold filter paper during a filtering operation. Funnels are made of heat-resistant glass, soft glass, or plastic. There are also several disposable types.

Graduated cylinders. GRADUATED CYLINDERS are tall, slender, cylindrical containers made of glass or plastic (Figure 3-6). They generally have a pour spout and a hexagonal base. They range in size from 10 to 4000 mL. Graduations are marked in 0.2-mL intervals on the 10-mL size, up to 50-mL intervals on the 4000-mL size. Graduated cylinders are used for measuring liquids quickly and without great accuracy.

Petri dishes. PETRI DISHES are shallow, vertical-sided dishes with flat bottoms. They usually have loose-fitting covers (Figure 3-7). They are used as containers for culturing standard plate counts and membrane filters. They can be glass or plastic and must be completely transparent for optimum visibility of colonies.

Usually, 100-mm × 15-mm size dishes are used for standard plate counts. A 50-mm × 12-mm petri dish with a tight bottom lid is used to contain and culture membrane filters. The tight fit retards evaporation loss from both broth and agar media, which helps maintain humidity inside the dish.

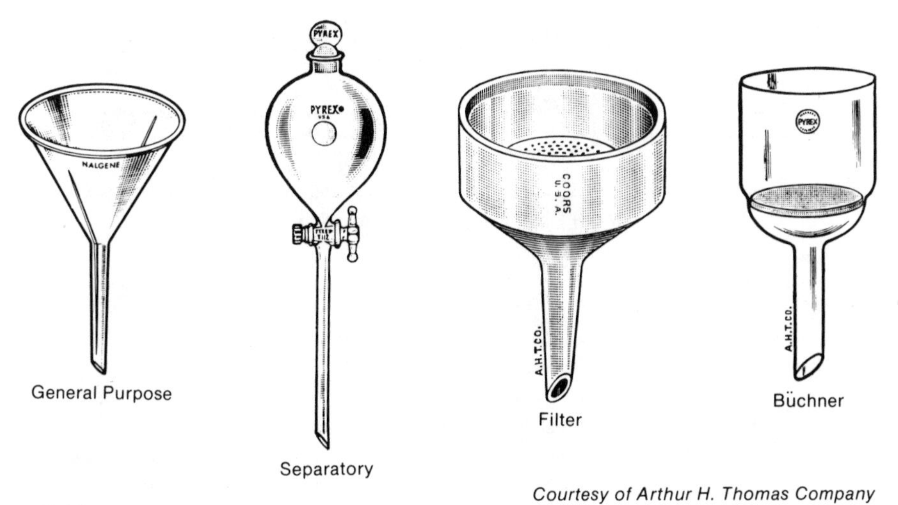

Courtesy of Arthur H. Thomas Company

Figure 3-5. Funnels

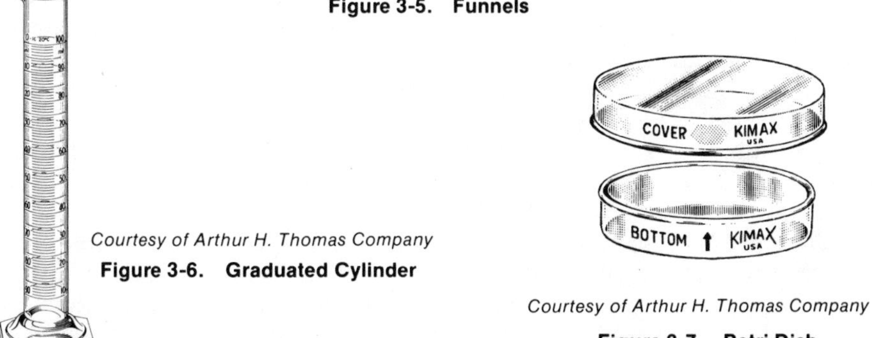

Courtesy of Arthur H. Thomas Company

Figure 3-6. Graduated Cylinder

Courtesy of Arthur H. Thomas Company

Figure 3-7. Petri Dish

Pipets. There are two kinds of PIPETS generally used. Pipets with a graduated stem, called MOHR PIPETS, can be used to measure any volume up to the capacity of the pipet. Pipets with a single measuring ring near the top are called VOLUMETRIC or TRANSFER PIPETS. Typical mohr and volumetric pipets are shown in Figure 3-8.

Pipets marked with the letters TD are designed "to deliver" the calibrated volume of the pipet. They will deliver the specified amount if the following conditions are met:

- The pipet is clean
- The pipet is held in a near vertical position during delivery
- Drainage is allowed to continue 5 sec after the level of liquid in the tip appears constant
- Contact is made between the tip of the pipet and the receiving vessel.

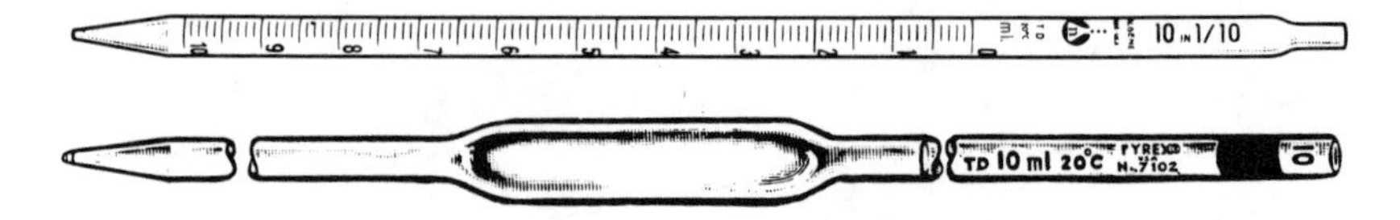

Courtesy of Arthur H. Thomas Company

Figure 3-8. Mohr Pipet (top) and Volumetric Pipet (bottom)

The small drop of solution that will be left in the pipet is accounted for in the pipet calibration. If a pipet has two bands ground into the glass at the top, it is calibrated for blowing out the last drop in the pipet.

Pipets are constructed with the delivery end tapered and with the opposite end fire polished so that it can be easily closed with a finger. For work requiring great accuracy, samples should be measured with a volumetric pipet. If the sample to be measured is less than 50 mL, it is good practice to use a pipet rather than a graduated cylinder. In general, transfer or volumetric pipets should be used when a great deal of accuracy is required. Measurement, or mohr, pipets may be used when great accuracy is not important.

Mouth suction should never be used to pipet solutions. A rubber suction bulb should be used in all cases.

Graduation marks on pipets must be legible and permanently bonded to the glass. Pipets should not stand overnight in caustic or detergent solutions because they may become cloudy or frosted. If pipets become badly etched, they should be discarded because poor visibility can interfere with accurate measurements.

Porcelain dishes. Porcelain labware has long been favored for use at elevated temperatures. Glazed porcelain is nonporous and highly resistant to heat, sudden changes in temperature, and chemical attack. Most EVAPORATING DISHES and FILTERING CRUCIBLES (called GOOCH CRUCIBLES) are made of porcelain. These dishes are used for total suspended solids analysis and total dissolved solids analysis. A porcelain evaporating dish and a filtering crucible are shown in Figure 3-9.

Reagent bottles. REAGENT BOTTLES (Figure 3-10) are made of borosilicate glass because they must be stable and resistant to heat and mechanical shock. The caps, also made of borosilicate glass, are usually ground-glass stoppers with flat tops, grip tops, or penny-head tops. Tops may also be plastic.

Reagent bottles should be used exclusively for storing reagents in the laboratory. They should be clearly labeled with the following information:

- Name of chemical and chemical formula
- Concentration
- Date prepared
- Initials of person who prepared the reagent
- Expiration date.

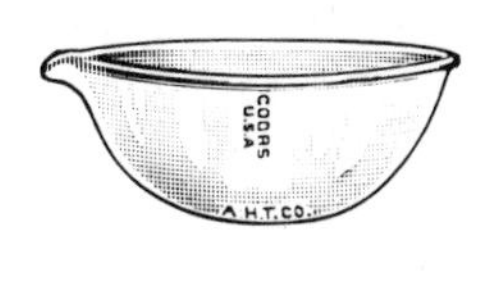

Courtesy of Arthur H. Thomas Company

Figure 3-9. Evaporating Dish (left) and Gooch Crucible (right)

Courtesy of Arthur H. Thomas Company

Figure 3-10. Reagent Bottles

Courtesy of Arthur H. Thomas Company

Figure 3-11. Sample Bottle

Courtesy of Arthur H. Thomas Company

Figure 3-12. Test Tube

Courtesy of Arthur H. Thomas Company

Figure 3-13. Culture Tube

Some reagent bottles are supplied with etched or raised glass letters; others have a specially ground area for marking.

Sample bottles. Wide-mouth SAMPLE BOTTLES are used for water sample collection, primarily because they are easier and quicker to fill than narrow-mouth bottles. In bacteriological sampling, there is less chance of contamination by splashing if wide-mouth bottles are used. Glass sample bottles should be made of borosilicate or corrosion-resistant glass, with metal or plastic closures equipped with nontoxic, leakproof liners (Figure 3-11).

Plastic bottles for bacteriological and chemical samples offer advantages of being low cost, lightweight, breakage resistant, and, depending on type, disposable.

Test tubes and culture tubes. TEST TUBES are hollow, slender, glass tubes with rounded bottoms and open tops (Figure 3-12). They generally have flared lips; CULTURE TUBES have plain lips (Figure 3-13).

Test tubes can be used for a variety of general laboratory tests. They can be made of disposable plastic, disposable glass, heat-resistant glass, or special-purpose glass.

Culture tubes are used for several tests in the water laboratory including multiple-tube fermentation tests for bacteria, biochemical tests for bacterial identification, and stock culture collections.

Cleaning Labware

It is important to clean labware as soon as possible after use. This will ensure an adequate supply of clean labware, and will promote cleaner labware by avoiding the formation of stains. Pipets and burets, for example, should be rinsed promptly after use. Use tap water first and then distilled water. Good labware cleaning procedure involves two washes and two rinses:

- Detergent wash
- Acid wash with 10-percent HCl
- Hot tap water rinse
- Distilled water rinse.

Any good household detergent is adequate for cleaning most laboratory glassware. Special detergents are also available from laboratory supply outlets.

Do not allow dissolved matter to dry on labware; future tests may be contaminated if labware is not cleaned promptly after use. If stubborn stains or crusty chemical residues remain after normal cleaning procedures, glassware should first be washed with a cleaning acid such as chromic acid. (Glassware to be used for chromium or manganese analyses should not be cleaned with chromic acid.) Chromic acid for cleaning is made by slowly adding 1 L of concentrated sulfuric acid (H_2SO_4) to 35 mL of saturated sodium dichromate solution, while stirring. The saturated sodium dichromate solution is prepared by adding sodium dichromate to distilled water. Dichromate is added until a residue forms on the bottom of the flask and will not dissolve. As moisture is absorbed from the air or from wet chemicals, the cleaning acid will lose its cleaning power. The mixture will eventually turn green and should be discarded. Glassware items with especially stubborn dirt films may be cleaned by soaking in chromic acid or organic acid detergents overnight. A typical method utilizes a 10-percent solution of an organic acid detergent.

Plastic bottles, plastic stoppers, and hard-rubber items can be destroyed if washed in chromic acid. Instead, concentrated hydrochloric acid (HCl) should be used.

3-2. Major Laboratory Equipment

All laboratories have several pieces of major laboratory equipment used to conduct laboratory tests. This section will acquaint you with the purpose and use of colony counters, desiccators, fume hoods, incubators, jar test apparatus, membrane filter apparatus, ovens, refrigerators, safety equipment (eye washes, deluge showers, fire extinguishers, and safety goggles), and water stills and demineralizers.

Colony counters. COLONY COUNTERS (Figure 3-14) are used to count bacterial colonies for the standard plate count test. Commercially manufactured colony counters magnify and backlight petri dishes so that bacterial colonies grown in the dishes can be identified and counted. Colony counters generally

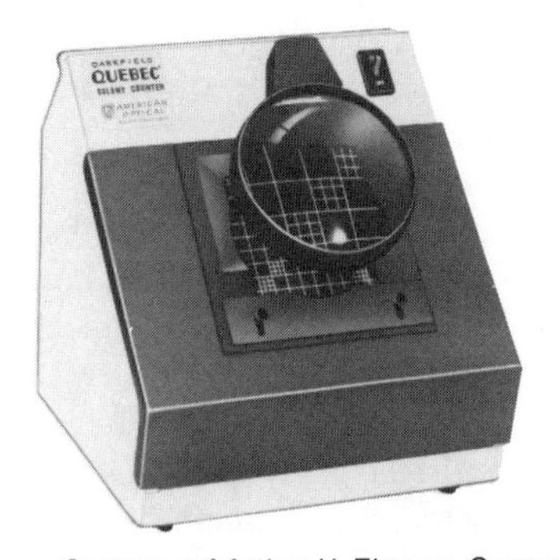

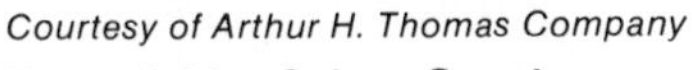
Courtesy of Arthur H. Thomas Company

Figure 3-14. Colony Counter

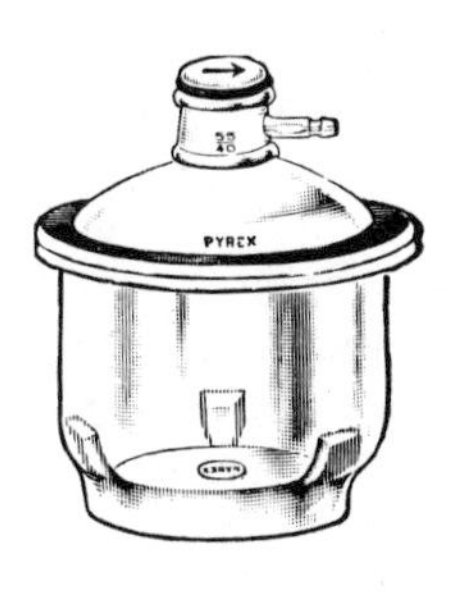

Courtesy of Arthur H. Thomas Company

Figure 3-15. Glass Desiccator

Courtesy of Arthur H. Thomas Company

Figure 3-16. Fume Hood

contain black contrast background with a ruled counting plate to make counting easier. The viewing area is illuminated from below the culture dish. The viewing field is magnified 1.5 times by a 5-in. (130-mm) lens (magnifying glass).

Desiccator. A DESICCATOR is a sealable container used to hold items before the items are weighed on an analytical balance. The desiccator serves two important functions: (1) it provides a place where heated items can cool slowly prior to weighing; (2) it provides a moisture-free environment so that items being cooled will not gain moisture weight before weighing. A chemical (such as dry calcium sulfate), placed in the bottom of the desiccator, removes moisture from the air within.

Glass desiccators with tight-fitting glass covers and ground-glass flanged closures (Figure 3-15) are the most popular type. Desiccating cabinets made of fiberglass or stainless steel and glass are also used by some larger laboratories.

Fume hoods. A FUME HOOD is a large, enclosed cabinet that contains a fan to vent fumes out of the laboratory. When used properly, this is one of the most important devices for preventing laboratory accidents. A typical fume hood (Figure 3-16) contains a glass or plexiglass door that can be closed to isolate the contents under the hood from the main laboratory. A convenient fume hood arrangement includes water drains, electrical outlets, gas taps, and vacuum and air pressure taps, all located within the fume hood cabinet.

Any laboratory test that produces unpleasant or harmful smoke, gas, vapors, or fumes should be conducted under a fume hood. Whenever heat is used in a test procedure, the test should be conducted under a fume hood. The hood contains the fumes; and the hood door, if partially lowered, can protect the operator's face and upper body from accidental splashing while performing the test.

Incubators. An INCUBATOR is an artificially heated container used in developing bacterial cultures for microbiological tests like those discussed in Module 4, Microbiological Tests. The three most common incubators are dry heat incubators, low temperature incubators, and water bath incubators.

Dry heat incubators contain a heating element capable of holding temperatures to within ±0.5° C of the desired incubation setting. They are useful for total coliform and standard plate count analyses that require a temperature of 35° C ± 0.5° C. These incubators usually have a temperature range of 30 to 60° C. Since they contain a heating element only, they cannot hold temperatures below room temperature. There are two types of dry heat incubators: gravity convection and forced air. The forced air incubators (Figure 3-17), have air circulating fans that help keep a constant temperature throughout the interior and therefore are more effective in maintaining the temperature tolerance limits than are the convection incubators.

Low temperature incubators are used for incubation in temperature ranges of −10 to 50° C with a ±0.3° C uniformity. These incubators are refrigerators that

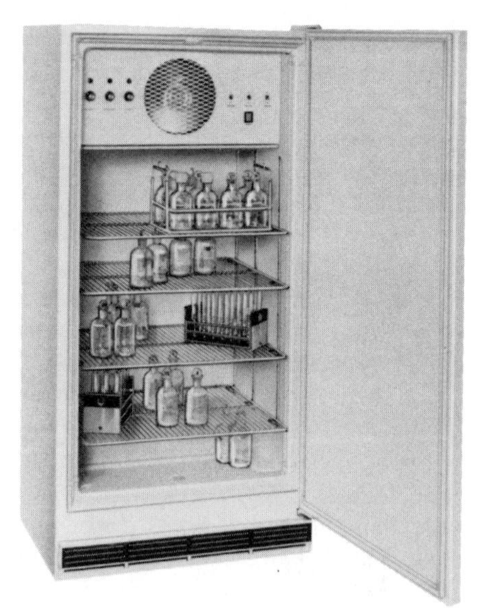

Courtesy of Arthur H. Thomas Company

Figure 3-17. Forced Air Incubator

contain a heating element and thermostat. They are most frequently used for BOD (biochemical oxygen demand) determinations. Low temperature incubators are more expensive than dry heat incubators since they must have both heating and cooling elements.

Water bath incubators are used for obtaining a more constant incubation temperature than can be maintained with a dry heat incubator. They are also used for many common analyses in which it is necessary to complete reactions with reagents or mixtures at a specified temperature. Water baths used as incubators for fecal coliform analyses must maintain a constant temperature of 44° C ±0.2° C. Water baths are capable of holding a ±0.2° C variation from the desired setting if the bath is covered and the water is circulated or gently agitated.

Most standard water baths are equipped with only heating elements to control temperature. These units operate in a range from room temperature to 100° C. There are some water baths available that have both refrigeration and heating elements and an operating range of from 0 to 100° C.

Jar test apparatus. A JAR TEST APPARATUS is an automatic stirring machine equipped with three to six stirring paddles and a variable-speed motor drive. The stirring machine is mounted on top of a floc illuminator, as shown in Figure 3-18. The illuminator provides the light needed for a clear visual inspection of the floc produced during the jar test. Use of the jar test apparatus is discussed in greater detail in Module 5, Physical/Chemical Tests.

Membrane filter apparatus. A MEMBRANE FILTER is capable of filtering particles as small as 0.45 μm from water. A typical apparatus consists of three basic parts: a filter holder base, a membrane filter, and a filter funnel. The apparatus fits on top of a vacuum filter flask (Figure 3-19) or on a suitably designed vacuum manifold (Figure 3-20). The filter holder base is available in stainless steel and fritted glass. The funnel is available in pyrex, plastic, and stainless steel.

Figure 3-18. Jar Test Apparatus

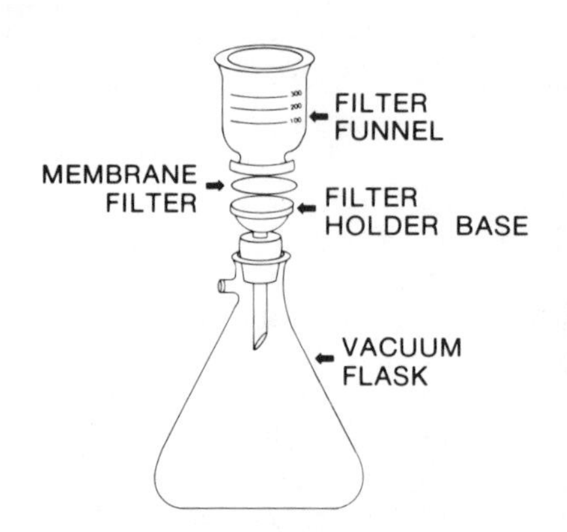

Figure 3-19. Membrane Filter Apparatus on Top of a Vacuum Filter Flask

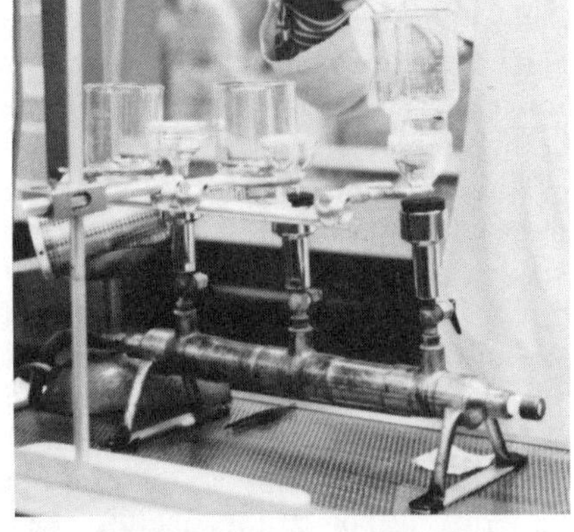

Figure 3-20. Membrane Filter Apparatus on Top of a Vacuum Manifold

The membrane filter apparatus, widely used in coliform bacteria analysis, can be used in many tests requiring small particle filtration. Detailed information on how to use the membrane filter apparatus is covered in Module 4, Microbiological Tests.

Ovens. OVENS are used primarily to dry, burn, or sterilize. The most commonly used ovens are utility ovens, muffle furnaces, and autoclaves.

UTILITY OVENS (Figure 3-21) have an operating temperature range from 30 to 350° C. They can be either gravity convection or forced air. In addition, some models are constructed so that a vacuum can be applied. These ovens are used for drying samples and labware (at 105° C) prior to weighing or sterilizing labware (at 170° C) for use in bacteriological testing.

MUFFLE FURNACES (Figure 3-22) are high-temperature ovens used to ignite or burn solids. The weight of the volatile materials is found by subtracting the

Courtesy of Arthur H. Thomas Company

Figure 3-21. Utility Oven

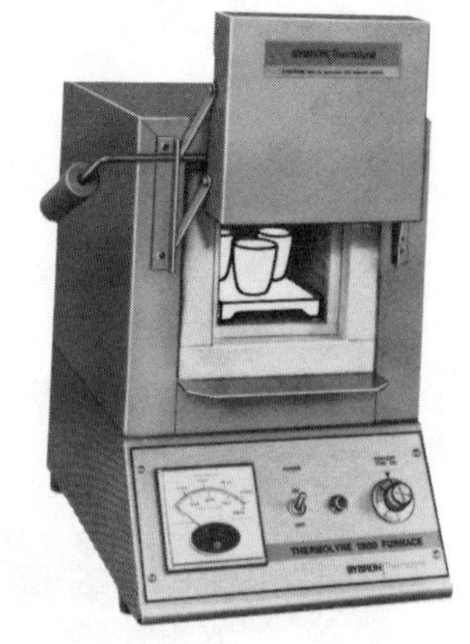

Courtesy of Arthur H. Thomas Company

Figure 3-22. Muffle Furnace

Courtesy of Arthur H. Thomas Company

Figure 3-23. Autoclave

weight after ignition from the weight before ignition. Muffle furnaces are lined with firebrick and generally have small ignition chambers. They usually operate at temperatures near 600° C.

AUTOCLAVES are used to sterilize such items as glassware, sample bottles, membrane filter equipment, culture media, and contaminated discard materials (Figure 3-23). They sterilize by exposing the material to steam at 121° C and 15 psi (100 kPa) for a specified period of time. Exposure time varies between autoclaves and with the amount and kind of material to be sterilized.

Refrigerators. REFRIGERATORS are used in laboratories to store chemical solutions and to preserve samples. There is a wide range of laboratory refrigerators available, but standard, home-type refrigerators are sufficient for most facilities.

Chemical solutions and samples should not be stored in the same refrigerator. Separate storage minimizes the chance of cross-contamination. Food should never be kept in a refrigerator that is used for sample or chemical storage.

Safety equipment. Two common laboratory hazards are chemical burns and fires. Every laboratory should be equipped to protect laboratory personnel from chemical burns and to extinguish small fires.

Safety goggles or protective face shields should be worn when there is danger of flying particles or spattering liquids. Although prescription glasses can be purchased with shatter-proof lenses, they do not surround the eyes with a tight covering to protect against splashes nor do regular safety glasses. The chemical SPLASH GOGGLES (Figure 3-24) or the FULL-FACE SHIELD (Figure 3-25) are specifically designed to reduce the chance of liquids reaching the eye. The lens material is also resistant to impact and penetration. Both types of eye protectors can be worn over normal prescription glasses.

NOTE: Contact lenses can increase injury from chemical splashes and should never be worn in laboratories or areas where such dangers exist, even if splash

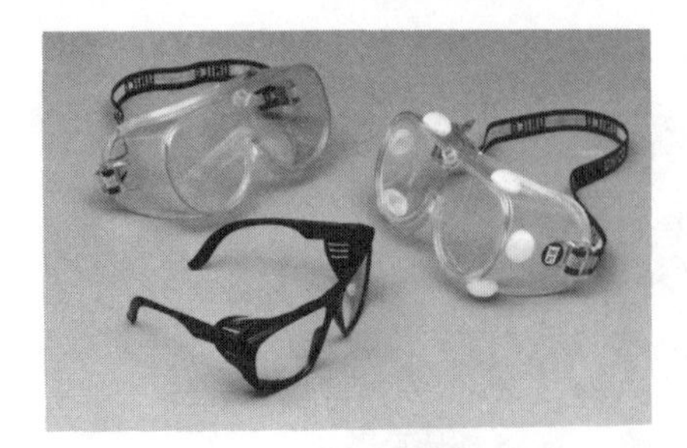

Courtesy of Bel-Art Products

Figure 3-24. Splash Goggles

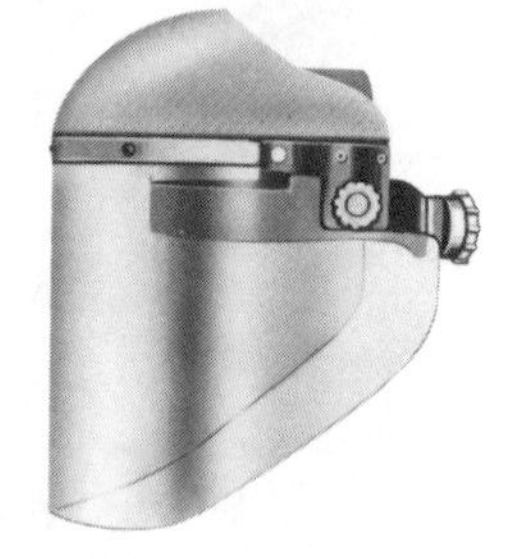

Courtesy of Arthur H. Thomas Company

Figure 3-25. Full-Face Shield

goggles or a face shield are worn. Permanent eye injury may result if even relatively mild chemicals are splashed into the eye and trapped between the contact lens and the eye's surface.

Eye washes should be available in every laboratory. Eyes are the most vulnerable part of the human body and should be protected. When a highly alkaline or acidic chemical touches the eyes or skin, deterioration begins immediately; the longer the period of contact, the more damage that will occur. The EYE WASH floods the eye with water as quickly as possible. Eye washes can be bottles with an eye cup or spray nozzle and 1-L reservoir used to flood the eye (Figure 3-26), or they can be permanent plumbing fixtures similar to a drinking fountain (Figure 3-27).

Deluge showers deliver a torrent of water in a uniform pattern to wash the body as completely and as rapidly as possible. As shown in Figure 3-28, a free-standing DELUGE SHOWER can be placed in a convenient, easy-to-reach location in the laboratory. The shower should have a large, easy-to-grab pull-chain ring or a paddle valve. Once the shower is turned on, it should remain on until turned off by a separate valve.

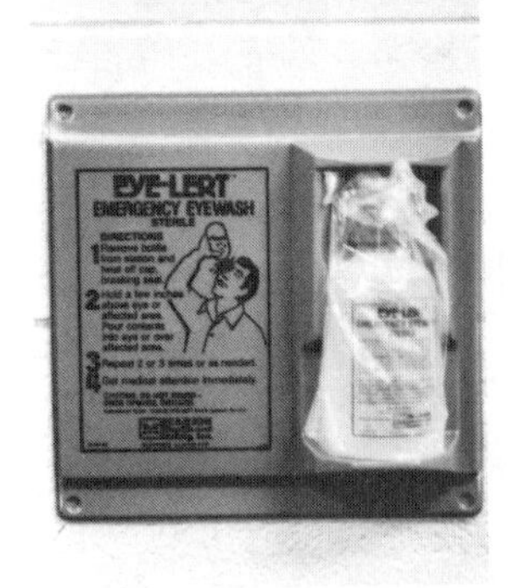

Figure 3-26. Eye Wash Bottle

Figure 3-27. Eye Wash Similar to a Drinking Fountain

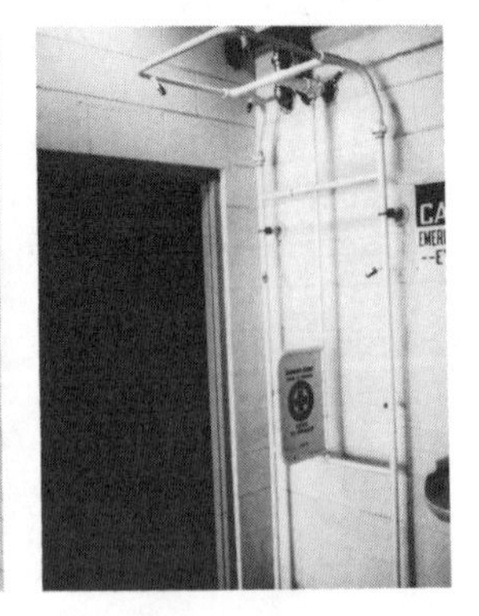

Figure 3-28. Deluge Shower

Fire extinguishers may prevent a large laboratory fire, if used quickly to put out a small fire. Each laboratory should have at least one all-purpose fire extinguisher (Figure 3-29) capable of putting out small fires. Laboratories should also be equipped with a fire blanket. Its major purpose is to extinguish burning clothing, but it can be used to smother liquid fires in small, open containers. The blanket is usually stored in a container (Figure 3-30) mounted on a wall or column and arranged in the container so that it can be easily pulled out. The fire extinguisher and fire blanket can extinguish almost all small fires that commonly occur in the laboratory.

Courtesy of Arthur H. Thomas Company

Figure 3-29. Fire Extinguisher

Courtesy of Arthur H. Thomas Company

Figure 3-30. Fire Blanket Stored in Container

Water stills and deionizers. There are two types of high purity water commonly used in most laboratories:

- Distilled water
- Deionized water.

A WATER STILL produces the distilled water needed for many laboratory tests and for rinsing all labware prior to use. Stills, like the one shown in Figure 3-31, take common tap water and, by evaporation and condensation, produce distilled water. Distilled water is free from dissolved minerals, uncombined gases, and all kinds of organic and inorganic nonvolatile contaminants. Stills can be portable or fixed, made of glass or of tin-coated metal; they can be heated by gas, electricity, or steam. Capacities range from about 0.3 to 5 gph.

A DEIONIZER removes all dissolved inorganic material (ions). Organic matter, uncombined gases, and fine particulates are not removed. Ion removal is accomplished by special ION EXCHANGE RESINS. As shown in Figure 3-32, deionizers are available in simple cartridge form for direct connection to any laboratory water faucet. The units continue to produce deionized water until the resin becomes exhausted. Exhaustion is signalled by a change in resin color or by

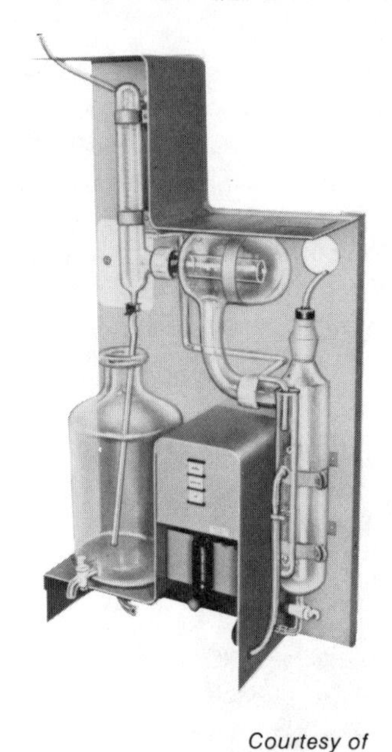

Courtesy of Arthur H. Thomas Company

Figure 3-31. Water Still

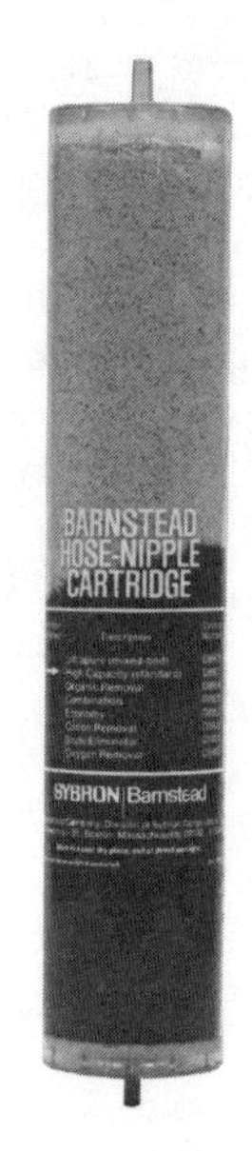

Courtesy of Barnstead Company

Figure 3-32. Deionizer

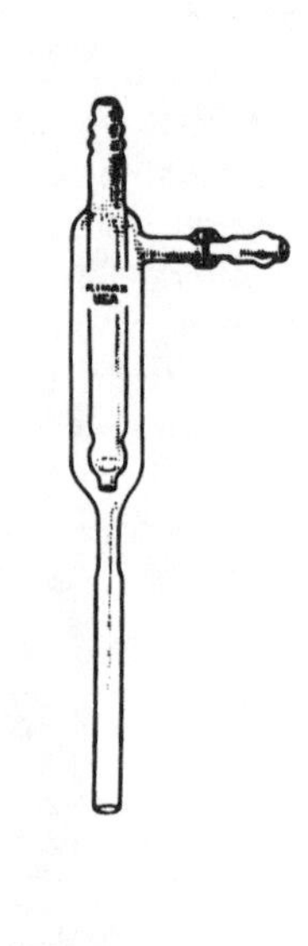

Courtesy of Arthur H. Thomas Company

Figure 3-33. Aspirator

a light or meter provided with the deionizer. When exhuasted, the old cartridge is removed and a new one inserted in its place. Deionizers have capacities up to 60 gph.

Deionized water can be used for most general laboratory purposes, including the preparation of solutions, washing of precipitates, extraction, and rinsing of glassware. Deionized water cannot be substituted for distilled water where organic impurities will interfere with an analytical method.

3-3. Support Laboratory Equipment

The well-equipped laboratory has a variety of support equipment used for various tests. You should be familiar with these items and some of their important uses. This section describes aspirators, hot plates, burners, filters, magnetic stirrers, and vacuum pumps.

Aspirators. An ASPIRATOR, a T-shaped plumbing fixture that connects to a water faucet, is used to create a vacuum. A typical aspirator is shown in Figure 3-33. When the faucet is turned on, water rushes down the vertical leg of the aspirator, creating a negative pressure (vacuum) in the horizontal stem. When connected to a vacuum filter flask, an aspirator produces the vacuum needed for many laboratory filtering operations.

Aspirators can create a cross connection and a potential hazard to the laboratory water supply. It is highly recommended that the faucet used be provided with an atmospheric vacuum breaker.

Hot plates. The HOT PLATE (Figure 3-34) is widely used for heating solutions. Hot plates have a temperature range from 100 to 500° C. The heating surface is smooth and solid for easy cleaning, and is made of a corrosion-resistant material such as glass, ceramic, or aluminum.

Burners. A gas BURNER is a convenient high-temperature heating device used in any laboratory served by natural gas or equipped with bottled gas (Figure 3-35). It is provided with adjustable air shutters for proper air-and-gas mixing.

Courtesy of Arthur H. Thomas Company

Figure 3-34. Hot Plate

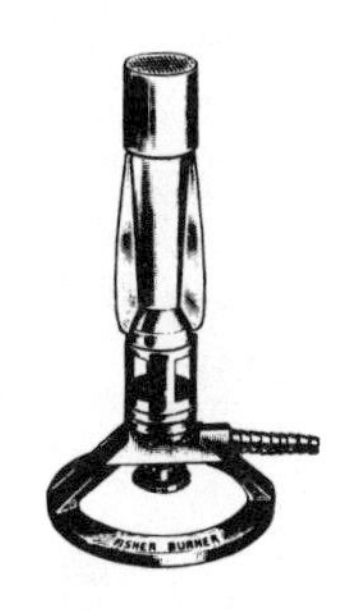

Courtesy of Arthur H. Thomas Company

Figure 3-35. Burner

Filters. Three types of FILTERS commonly used in most laboratories today are filter paper, glass-fiber filters, and membrane filters.

FILTER PAPER is used to clarify solutions, collect particulates, and separate solids from liquids. The paper's pore size is between 5 and 10 μm and the filters are available in diameters ranging from 4.25 to 50 cm.

GLASS-FIBER FILTERS are made of uniform, fine glass fibers. They are used to filter fine particulates, bacteria, and algae while retaining a high rate of flow. Pore size varies from 0.7 to 2.7 μm and diameter varies from 15 to 261 cm.

MEMBRANE FILTERS are cellulose acetate membranes with precise pore size ranging from 0.2 to 5.0 μm. They are available in diameters from 13 to 142 mm. They have many uses in the laboratory; however, the main use is for bacterial testing.

Magnetic stirrers. MAGNETIC STIRRERS continuously stir solutions for long periods of time. They are similar in appearance to laboratory hot plates. The surface has a corrosion-resistant top of aluminum, glass, or ceramic, which covers a variable-speed, rotating magnetic field. This magnetic field spins a magnetized, PTFE plastic-coated stirring bar. Combination magnetic stirrer-hot plate units contain separately controlled heating elements and stirring mechanisms. The units can function as heaters, stirrers, or both.

Vacuum pumps. VACUUM PUMPS (Figure 3-36) are useful for a variety of laboratory analyses. The most common use is to aid in filtration. Large, well-equipped laboratories may have a large vacuum pump connected by pipes to taps in different areas of the laboratory. However, most smaller laboratories

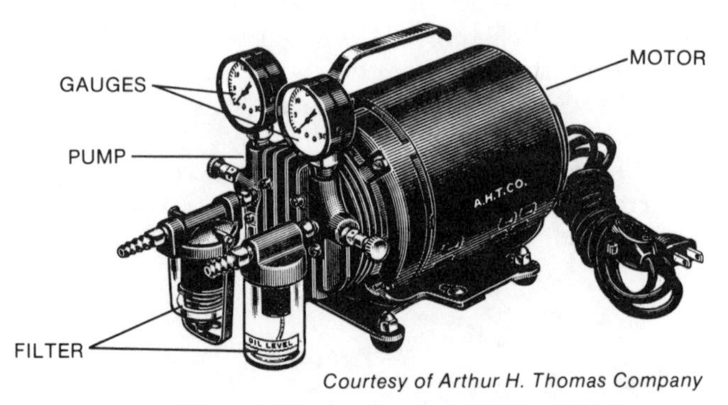

Courtesy of Arthur H. Thomas Company

Figure 3-36. Pressure-and-Vacuum Pump

find a small, portable, pressure-and-vacuum pump suitable. These portable units typically use a 1/8- to 1/3-hp electric motor, and can produce vacuums up to 28 in. of mercury (94 kPa) and pressures up to 50 psig (350 kPa, gauge). The attached filters, shown in Figure 3-36, help to provide oil-free output air.

3-4. Laboratory Instruments

Many instruments in a water laboratory are sensitive and accurate devices used to measure a water characteristic. There is a wide variety of laboratory instruments, varying from the simplest thermometer to the very sophisticated atomic absorption spectrophotometer (AA unit). This section describes two types of balances, several meters, and optical microscopes.

Balance. The BALANCE, a delicate instrument used to measure weight, is one of the most important laboratory tools and must be handled carefully. Directions in an instruction or operating manual for each type of balance should always be followed to ensure proper use of the specific balance and achievement of correct results.

The pan balance, or "rough" balance, weighs loads up to approximately 2 kg (about 4.5 lb). Pan balances are available in single- and double-pan models, as shown in Figure 3-37.

To use the SINGLE-PAN BALANCE, the item to be weighed is placed on the pan and the counterweights, located on the three horizontal arms (beams), are adjusted. The indicator arrow on the far right end of the three beams will show when the counterweights equal the weight of the item. The weight reading is obtained by adding the amount of weight shown on each of the three beams.

The DOUBLE-PAN BALANCE has the same capability as the single-pan; however, the procedure is more time consuming. Standard brass weights must be placed on the right-hand pan until the two pans can be balanced by turning the knob or sliding a weight across a beam on some models. The weight of the item is found by adding the amount of standard brass weight used to the dial reading (beam reading).

ANALYTICAL BALANCES, far more sensitive and precise than pan balances, can detect a change in weight as little as ±0.0001 g (0.1 mg). The most convenient and

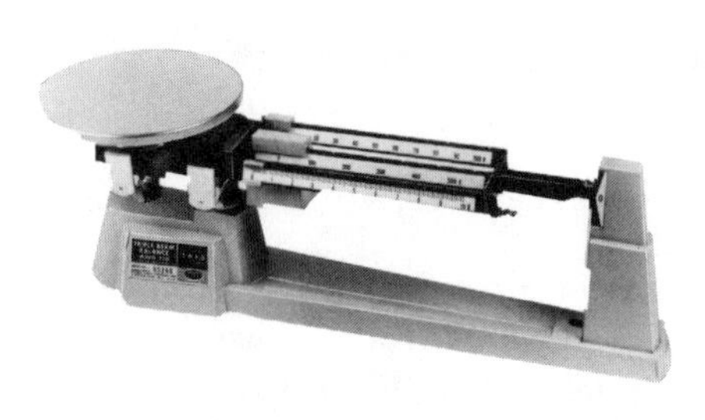

Courtesy of Ohaus Scale Corp.

Figure 3-37. Single- and Double-Pan Balances

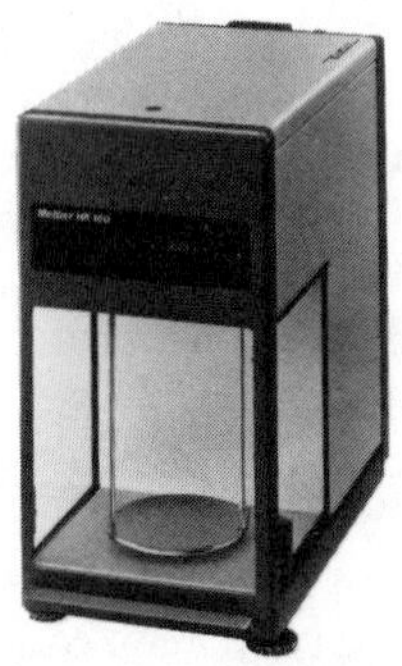

Courtesy of Arthur H. Thomas Company

Figure 3-38. Automatic Single-Pan Analytical Balance

Courtesy of Arthur H. Thomas Company

Figure 3-39. Marble Balance Table

practical analytical balance used in today's laboratory is the automatic single-pan balance (Figure 3-38). The word "automatic" refers to the built-in standard weights, which are placed into operation quickly and "automatically" by simply turning a knob. The final weight is easily read from a display. This system simplifies the weighing procedure and minimizes errors in recording weights. Rapid weighing is an important feature of the analytical balance since items being weighed can pick up moisture from the air, causing a slow weight gain.

A balance must be located on a solid, level surface for it to function properly. Metal counters, for example, are not suitable because of flexibility and movement in the metal. Vibrations from nearby machinery or from an unstable floor can be transmitted through the table to the balance causing an inaccurate reading. The vibrations from surrounding machinery can be greatly reduced if a properly constructed table is located on an unyielding floor.

Most laboratories use solid marble balance tables like the one shown in Figure 3-39. These tables reduce transmission of vibrations from the floor, and remain level. They should be placed on concrete floors (slab-on-grade at ground level or near a bearing wall if above ground level) and in a relatively constant environment.

Meters. A variety of specialized METERS are used in water treatment plant laboratories for measuring water quality characteristics.

COLORIMETERS measure the concentration of a constituent by measuring the intensity of a color. Colorimetric measurements may be made using a wide variety of equipment including standard color comparison tubes, photoelectric colorimeters, and spectral photometers. Each has its place and particular application in the laboratory.

COLOR COMPARATORS with permanent color standards for specific parameters can be purchased for laboratory and field use. There are two types of comparators. The disk type consists of a wheel of small colored glasses. The slide type consists of liquid standards and glass ampules. These comparators give rapid, fairly acceptable, consistent results. The most common comparators are the chlorine residual test kit, used by most water utilities, and the chlorine-pH test kit, used for swimming pools.

The PHOTOMETER is an electronic device that performs the same function as a colorimeter or color comparator. Far more accurate and precise than visual colorimeters, the photometer can measure small differences in color intensities not easily seen by the naked eye. Other advantages over visual colorimeters include freedom from variable light conditions and elimination of errors due to color blindness or color bias of the analyst.

The photometer is versatile, easy-to-use, and relatively inexpensive. The USEPA drinking water regulations allow the use of photometers in testing for nitrate, arsenic, fluoride, and chlorine residual.

A basic photometer, such as the one in Figure 3-40, has five main components: (1) white light source, (2) wavelength control unit, (3) sample compartment, (4) detector, and (5) meter. The white light passes through the wavelength control unit (a simple colored filter, a diffraction grating, or a prism) to produce a single-color light (light of a specific wavelength). The single-color light then passes through the treated sample, which is contained in a glass tube (called a cuvette) in the sample compartment. The amount of light that passes through the sample is sensed by the detector and indicated on the graduated-scale meter. The

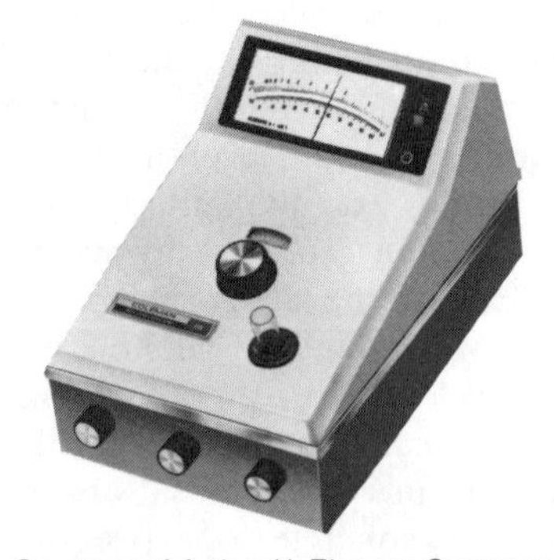

Courtesy of Arthur H. Thomas Company

Figure 3-40. Photometer

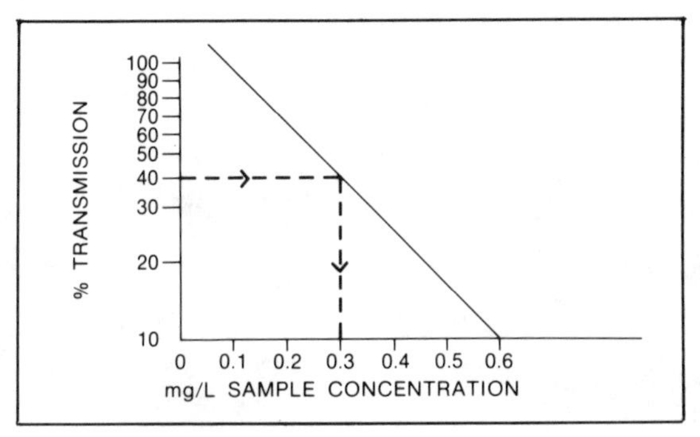

Figure 3-41. Calibration Curve

measurement can be expressed either in terms of percent transmittance or in terms of absorbance. Finally, the concentration of the measured constituent is found from a previously prepared calibration curve (Figure 3-41). Such a curve must be prepared for each constituent to be measured. Once prepared, the curve should be valid for several years.

There are two basic types of photometers, the ELECTROPHOTOMETER and the SPECTROPHOTOMETER. The basic difference between the two is the method used to produce the single-color light.

An electrophotometer uses a simple-colored glass filter. A specific filter color is required for each constituent measured. Usually, electrophotometers are used for just a few difficult constituent determinations.

A spectrophotometer uses either a diffraction grating or a prism to control the light color. By adjusting the angle of the grating or the prism, different light colors (different wavelengths of light) can be selected. Thus, one adjustable grating or prism provides a continuous spectrum of color selections. This is not possible with the electrophotometer. A spectrophotometer is particularly suited when a wide variety of constituents are being measured. Its versatility allows convenient selection of the best light color for any test.

A special type of spectrophotometer, the ATOMIC ABSORPTION SPECTROPHOTOMETER (AA unit), is used for analyses of most heavy metals in water. It is a sophisticated and expensive analytical tool that must be operated by specially trained laboratory technicians.

A pH METER is a sensitive voltmeter that measures the pH of samples. There are many instrument types available; a typical meter is shown in Figure 3-42. The meter scale is graduated in pH units from 0 to 14. More sophisticated meters have expanded scales that allow more precise pH measurement within a narrower range and a millivolt scale that allows measurement of specific ions such as fluoride. Two electrodes are supplied with the meter. One electrode, a standard calomel reference type, develops a constant voltage to compare against the changing voltage of the second. Voltage of the second electrode, a glass type, changes as pH changes. The second electrode is designed so that a change of one pH unit produces a voltage change of 59.1 mV at 25° C. In some units, the two

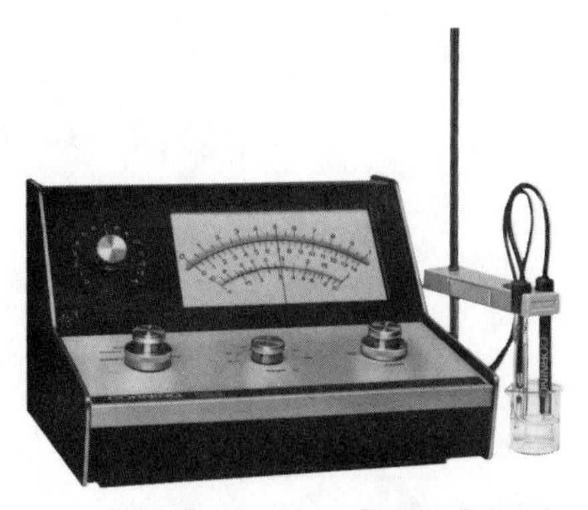

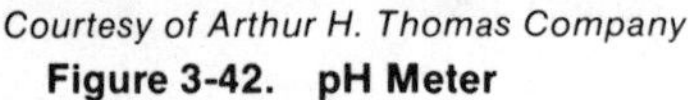

Courtesy of Arthur H. Thomas Company

Figure 3-42. pH Meter

Courtesy of Orion Research Inc.

Figure 3-43. Specific Ion Meter

electrodes are mounted in a single unit, called a combined electrode. The meter's control switches include an on-off switch, temperature compensation, and a standardizing adjustment knob.

SPECIFIC ION METERS measure the concentration of a specific constituent in water, such as fluoride. The complete unit consists of a millivolt meter and interchangeable electrodes. Each electrode is selectively sensitive to one particular constituent of the water, and each specific ion test requires a different electrode. There are currently more than 20 selective electrodes, including electrodes that will measure chloride, copper, hardness, fluoride, sodium, and chlorine.

In general, most specific ion electrodes are only useful for applications where many consecutive tests must be made on similar samples. Frequent calibrations may be necessary, which are often more time consuming than testing by other methods. Also, the electrodes are subject to interferences. The fluoride electrode is an exception; the results obtained are excellent. Only high pH values interfere with fluoride test results, and since pH can be lowered before testing, the interference can be easily eliminated.

The meter used with the electrodes (Figure 3-43) resembles a pH meter. There are two major differences: the addition of a millivolt scale on the meter face and the provision for use of selective ion electrodes. Often, a pH meter is purchased with a millivolt scale so that it can also be used as a specific ion meter. A specific ion meter may read concentration directly, or it may read in millivolts, in which case, concentration is determined by using a standard curve. When using a meter with millivolt readings, a standard curve to convert from millivolts to concentration must be developed. This is done by measuring several samples of known concentration and plotting the results.

A TURBIDIMETER measures the clarity of water. Basically, the amount of light impeded by or scattered by the suspended particles in the sample of water is measured. The USEPA drinking water regulations specify the nephelometric method as the only approved method for measuring turbidity. Although there are other methods, the nephelometric method will be the only one discussed.

NEPHELOMETRIC TURBIDIMETERS are very similar to photometers, both in appearance and in performance. The turbidimeter consists of the following major components:

- Light source
- Focusing device
- Sample compartment
- Detector (photomultiplier tube)
- Meter.

As shown by Figure 3-44, the light passes through a focusing device, into the sample compartment, passing through the sample. The light is reflected by the individual particles that cause turbidity. Some of that reflected light strikes a detector, such as a phototube, located 90 degrees off the main light path, which measures the amount of light reaching it. The meter indicates the corresponding turbidity in NEPHELOMETRIC TURBIDITY UNITS (NTU).

Nephelometric turbidimeters, like those shown in Figure 3-45, are quick and relatively easy to standardize and operate. Most of them have meter read-outs that indicate turbidity values directly. The meters usually have several scale ranges. The most common ranges are 0–0.2 NTU, 0–1 NTU, 0–10 NTU, 0–100 NTU, and 0–1000 NTU.

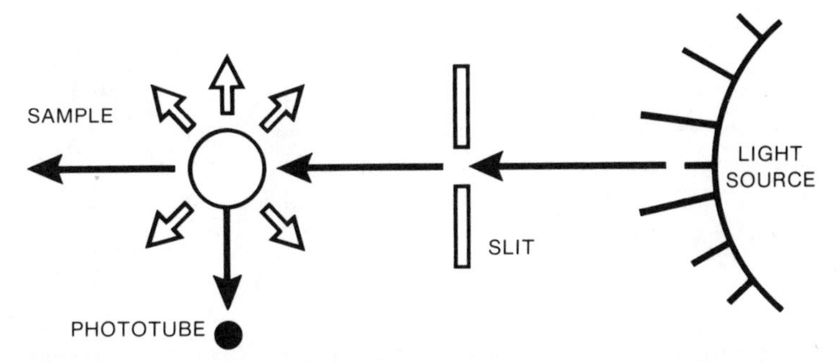

Figure 3-44. Path of Light Through a Nephelometric Turbidimeter

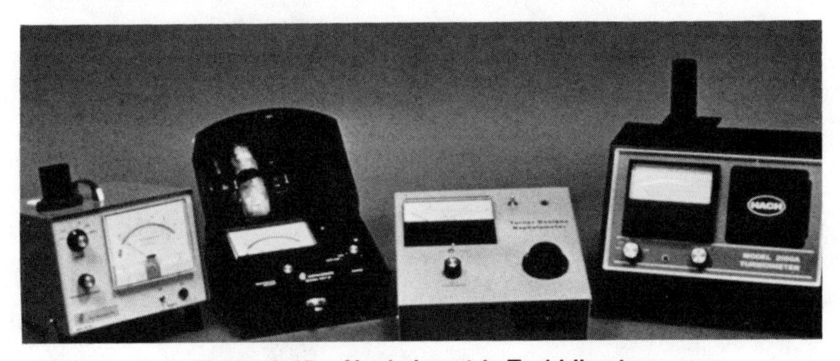

Figure 3-45. Nephelometric Turbidimeters

Since all communities using surface-water sources are required to test their treated water daily for turbidity, the turbidimeter is a necessity at every plant, and will be one of the most commonly used instruments.

A variation of the basic turbidimeter is the continuous monitoring on-line type. Instead of a sample compartment, it has a flow-through chamber in which turbidity is continuously measured. A complete on-line installation typically consists of the flow-through nephelometric sensor, a meter-type turbidity indicator, and a chart recorder. Such installations are used to monitor raw-water, in-process water, and finished-water turbidities.

Microscopes. A MICROSCOPE magnifies extremely small objects so that they can be seen and studied. The naked eye can see objects as small as 40 μm, about half the diameter of a human hair. Through very powerful magnification, the microscope extends human vision into the incredibly small worlds of algae, bacteria, and viruses.

The simplest optical microscope is a magnifying glass. The best magnifying glass can magnify an object 20 times. (Abbreviated 20×, meaning the diameter of the image is 20 times greater than the diameter of the object.) The most common microscope is the COMPOUND MICROSCOPE. A compound microscope uses two or more lenses instead of the one lens used in a magnifying glass. A compound microscope consists of five basic parts, shown in Figure 3-46.

- Stand.
- Movable stage.
- Head, including the oculars, or eyepiece lenses. A one-ocular head is called a monocular; two-ocular, binocular (shown); and three-ocular, triocular (one ocular is used for mounting a camera).
- Objective nosepiece, a revolving set of lenses. The unit shown has four different objective lenses. The selection of different objectives gives different magnifications.
- Illuminator (light source) and condenser lens used to focus light onto the object being viewed.

Ordinarily, the compound microscope magnifies to 1000 times, though some advanced types reach much higher magnification. As shown in Table 3-1, the

Table 3-1. Optical Microscope Magnification

Type of Objective	*Overall Magnification*	
	10x Ocular	*15x Ocular*
16 mm (10x or low power)	100x	150x
8 mm (20x or medium power)	200x	300x
4 mm (43x or high power)	430x	645x
1.8 mm (90x or oil immersion)	900x	1350x

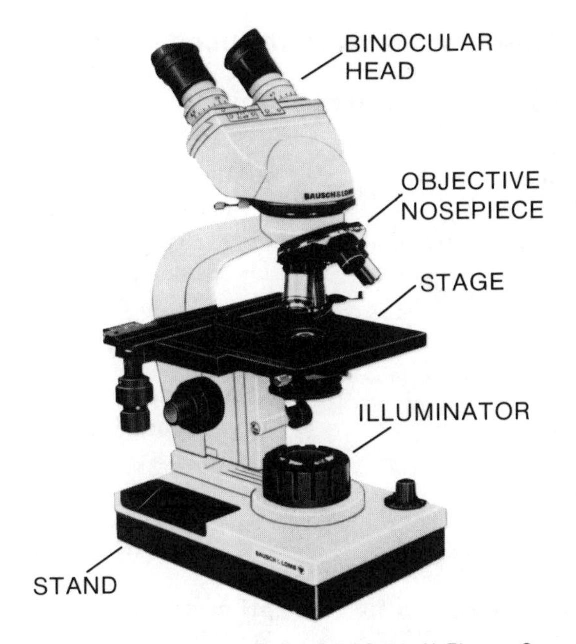

Courtesy of Arthur H. Thomas Company

Figure 3-46. Compound Microscope

magnifying power of a compound microscope depends on the combined magnification of the eyepiece and the objective lenses. For example, a 20× objective lens combined with a 10× eyepiece produces a magnification of 200× ($20 \times 10 = 200$).

The compound microscope is one of the most important tools in the water quality laboratory. It is used for counting and identifying the microscopic plant and animal life typically found in water, including color-, taste-, and odor-causing algae and disease-causing bacteria.

Selected Supplementary Readings

Simplified Procedures for Water Examination. AWWA Manual M12. AWWA. Denver, Colo. (1977).

Guidelines for the Selection of Laboratory Instruments. AWWA Manual M15. AWWA. Denver, Colo. (1978).

Glassware—The Lab's Best Friend. *OpFlow.* 2:1:1 (Jan. 1976).

Glossary Terms Introduced in Module 3

(Terms are defined in the Glossary at the back of the book.)

Analytical balance
Aspirator
Atomic absorption spectrophotometer
Autoclave
Autoclaved
BOD
Balance
Beaker
Borosilicate glass
Buret
Burner
Colony counter
Color comparator
Colorimeter
Compound microscope
Culture tube
Deionizer
Deluge shower
Desiccator
Dilution bottle
Double-pan balance
Electrophotometer
Erlenmeyer flask
Evaporating dish
Eye wash
Filter
Filter paper
Filtering crucible
Flask
French square (see dilution bottle)
Full-face shield
Fume hood
Funnel
Glass-fiber filter
Gooch crucible
Graduated cylinder
Hot plate
Incubator
Ion exchange resin
Jar test apparatus
Magnetic stirrer
Membrane filter
Meter
Microscope
Milk dilution bottle (see dilution bottle)
Mohr pipet
Muffle furnace
Nephelometric turbidimeter
Nephelometric turbidity unit (NTU)
Oven
pH meter
Petri dish
Photometer
Pipet
Reagent bottle
Refrigerator
Sample bottle
Single-pan balance
Specific ion meter
Spectrophotometer
Splash goggles
TD
Test tube
Titration
Transfer pipet (see volumetric pipet)
Turbidimeter
Utility oven
Vacuum pump
Volumetric flask
Volumetric pipet
Water still

Review Questions

(Answers to Review Questions are given at the back of the book.)

1. Describe each of the following and give the use and sizes (where given) for each:
   (a) Buret
   (b) Graduated cylinder
   (c) Petri dish
   (d) Dilution bottle
   (e) Culture tube.

2. List five pieces of information to be included on labels of reagent bottles.

3. What are the four steps of good labware cleaning?

4. Define major laboratory equipment.

5. Describe each of the following pieces of equipment and give its purpose:
   (a) Desiccator
   (b) Membrane filter apparatus
   (c) Utility oven
   (d) Deluge shower.

6. How do low temperature incubators differ from other types of incubators?

7. What substances are removed by water stills?

8. What hazard can an aspirator create?

9. What is the most common use for vacuum pumps?

10. List three types of filtering material.

11. Which would be the best balance for weighing (a) a beaker of alum (b) a small quantity of microbiological media?

12. Why should time be a factor in weighing accuracy?

13. Which type of photometer would be most useful for a lab that has to test 50 constituents a day? Why?

Study Problems and Exercises

1. You traditionally have not conducted process control tests at your water treatment plant. The new city manager has issued a directive to the water department requiring operators to start running tests for (a) turbidity, (b) coagulant effectiveness (jar test), (c) alkalinity, (d) pH, (e) temperature, and (f) chlorine residual on a routine basis. The manager has requested that you prepare a list of instruments and labware needed to conduct these tests. Prepare a report itemizing needed labware and instrument for each test that is to be conducted routinely.

Module 4

Microbiological Tests

A variety of different microorganisms are found in untreated water. Most of these organisms do not pose a health hazard to humans. The organisms that operators are concerned with most are those that cause disease (pathogenic organisms, or PATHOGENS), which include certain bacteria, virus, and protozoa. Table 4-1 lists the more common waterborne diseases and the organisms that cause the disease, as well as the disease symptoms.

Most organisms found in water are not pathogenic; however, even microorganisms that do not cause disease can create problems in a water plant or distribution system. Some of these organisms can cause taste and odor problems that are difficult to correct. Others may create corrosion problems in iron pipes of the distribution systems, causing red water and stained plumbing fixtures in consumers' homes.

Two common analytical methods to determine the MICROBIOLOGICAL quality of water are the TOTAL COLIFORM TEST and the STANDARD PLATE COUNT; both test for bacteria. This module will discuss these tests and their significance to water supply operations.

After completing this module you should be able to

- Describe the significance of pathogens in water.
- Explain why coliform bacteria are used as an indicator of pathogenic bacteria.
- Understand the significance and methods of determination for total coliform (multiple-tube and membrane filter methods) and standard plate count.

Table 4-1. Waterborne Diseases

Waterborne Disease	*Causative Organism*	*Source of Organism in Water*	*Symptom*
Gastroenteritis	*Salmonella* (bacteria)	Animal or human feces	Acute diarrhea and vomiting
Typhoid	*Salmonella typhosa* (bacteria)	Human feces	Inflamed intestine, enlarged spleen, high temperature—fatal
Dysentery	*Shigella* (bacteria)	Human feces	Diarrhea, rarely fatal
Cholera	*Vibrio comma* (bacteria)	Human feces	Vomiting, severe diarrhea, rapid dehydration, mineral loss—high mortality
Infectious hepatitis	Virus	Human feces, shell fish grown in polluted waters	Yellowed skin, enlarged liver, abdominal pain, low mortality, lasts up to 4 months
Amebic dysentery	*Entamoeba histolytica* (protozoa)	Human feces	Mild diarrhea, chronic dysentery
Giardiasis	*Giardia lamblia* (protozoa)	Suspect wild animal feces	Diarrhea, cramps, nausea and general weakness, lasts 1 week to 30 weeks—not fatal

4-1. Coliform Bacteria

Table 4-1 lists several diseases of pathogenic origin that are transmitted to humans through contaminated water. It is impractical to test water for every known pathogen; pathogens are very difficult to identify, and it is not economical to routinely test for specific types.

A more practical approach is to examine the water for INDICATOR ORGANISMS that are specifically identified with contamination. Indicator organisms should:

- Always be present in contaminated water
- Always be absent when contamination is not present
- Survive longer in water than pathogens
- Be easily identified.

The COLIFORM (also called TOTAL COLIFORM) group of bacteria meets all criteria for an ideal indicator. These bacteria are generally not pathogenic, yet they are usually present when pathogens are present. Additionally, coliform bacteria are more plentiful than pathogens and can often stay alive in the environment for longer periods of time.

Coliform bacteria are easily measured in the laboratory. As a rule, where coliforms are found to be present in water, it is assumed that pathogens may also be present, making the water bacteriologically unsafe to drink. If coliform bacteria are absent, the reverse is true.

Two methods exist for determining the number of coliform bacteria in a water sample: the MULTIPLE-TUBE FERMENTATION METHOD and the MEMBRANE FILTER METHOD. The detailed analytical procedures for these tests can be found in the latest edition of *Standard Methods for the Examination of Water and Wastewater*.

Significance

Coliform bacteria detected in a water sample with either multiple-tube fermentation or the membrane filter technique warns of possible contamination. One positive test does not conclusively prove contamination however, and additional tests must be conducted.

Samples may become contaminated from external sources such as improper sampling, improperly sterilized bottles, and laboratory error. Regulatory agencies recognize this fact, and drinking water regulations do not require check or repeat sampling after occasional findings of low levels of coliforms. However, continued detection of coliforms reveals that contamination is present and that the water is unsafe. (Drinking water regulations and MCLs for coliform are discussed in Module 1, Drinking Water Standards.)

Sampling

Sterile bottles must be used for all samples collected for bacteriological analyses. The same sampling procedures should be used for coliform analysis and standard plate count analysis. Refer to Module 2, Sample Collection, Preservation, and Storage for these procedures.

Test Methods

Multiple-tube fermentation. The multiple-tube fermentation test progresses through three distinct steps: the PRESUMPTIVE TEST, the CONFIRMED TEST, and the COMPLETED TEST. The confirmed test and the completed test increase the certainty that positive results obtained in the presumptive test are due to coliform bacteria and not other kinds of bacteria. The completed test is used to establish definitively the presence of coliform bacteria for quality control purposes. To check its procedures, the laboratory should conduct the completed test on at least 10 percent of the positive tubes from the confirmed test. Bacteriological testing of most public water supplies stops after the confirmed test. This is the

Figure 4-1. Typical Multiple-tube Setup

minimum test that must be applied to all samples when using the multiple-tube fermentation method. See Figure 4-1 for a typical multiple-tube setup.

The presumptive test is the first step of the analysis. Samples are poured into each of five tubes containing a culture media (lactose or lauryl tryptose broth) and an inverted vial. The samples are then INCUBATED for 24 hours, checked, incubated for 24 hours more, and checked again. If coliform bacteria are present in the water, gas will begin to form in the inverted vials within the 48-hour period; this indicates a POSITIVE sample. If no gas forms, the sample is NEGATIVE. If gas is produced after either the 24-hour or 48-hour incubation period, all positive samples undergo the confirmed test.

The confirmed test is more selective for coliform bacteria. Cultures from the positive samples in the presumptive test are transferred to brilliant green lactose bile broth and incubated. If no gas is produced after 48 hours of incubation, the test is negative and no coliform bacteria are present. If gas is produced, the test is positive indicating the presence of coliform bacteria. The number of positive tubes is used to determine compliance with the maximum contaminant level (MCL) of the USEPA drinking water regulations.

Positive samples then undergo the completed test. A sample from the positive confirmed test is placed on an eosin methylene blue (EMB) AGAR plate and incubated. Coliform colonies will form on the EMB plates. A small portion of the coliform colony is transferred to a nutrient agar slant and incubated for 18 to 24 hours. A second portion is transferred to a lauryl tryptose broth and incubated for 24 to 48 hours. The completed test is positive if (1) gas is produced in the lauryl tryptose broth and (2) red-stained, nonspore forming, rod-shaped bacteria are found. If no gas is produced in the lauryl tryptose broth or if no red-stained, chain-like cocci or blue-stained, rod-shaped bacteria are found on the agar, then the test is negative. Refer to Figure 4-2 for a summary of the multiple-tube fermentation method.

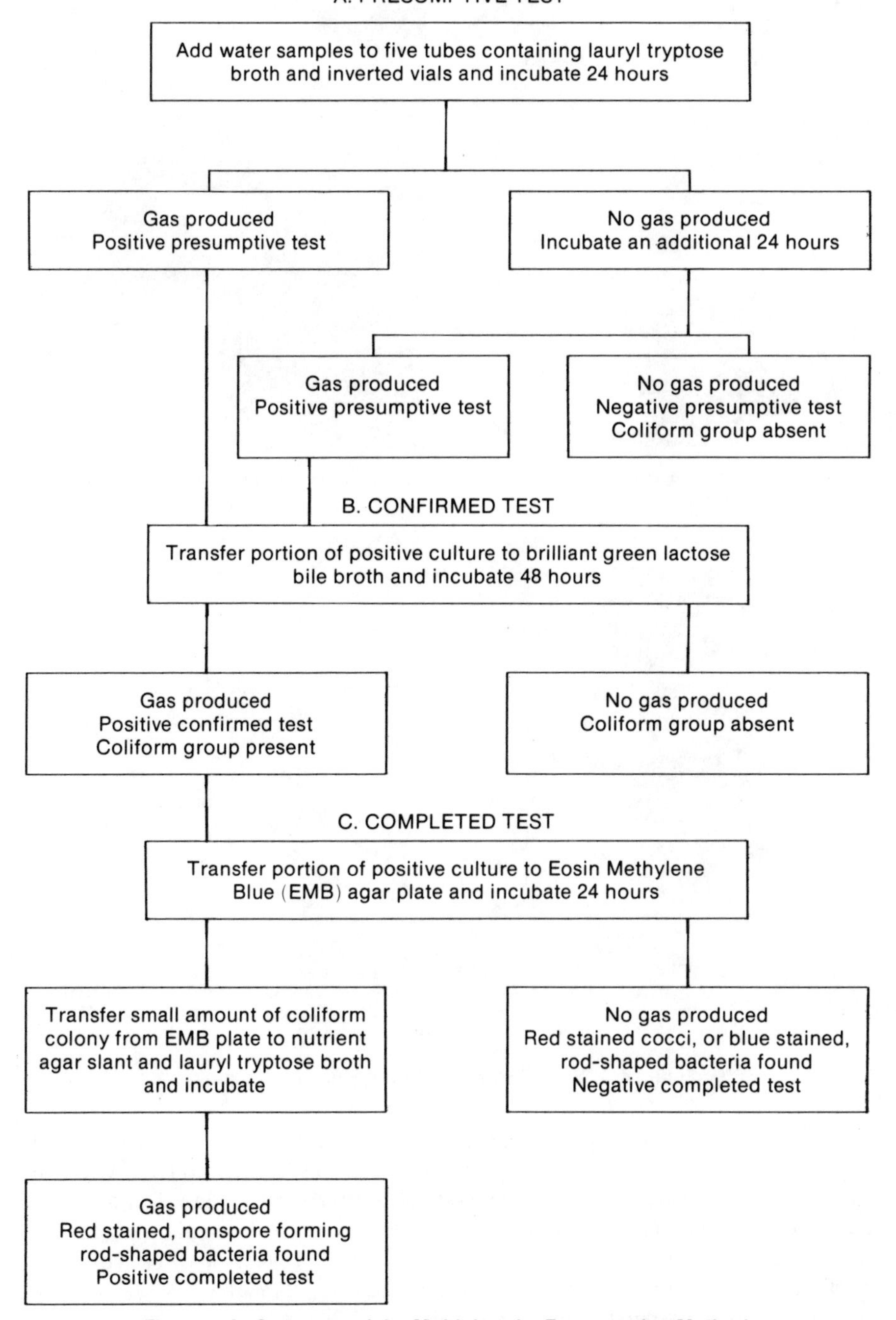

Figure 4-2. Summary of the Multiple-tube Fermentation Method

(a) Sample being poured through a filter

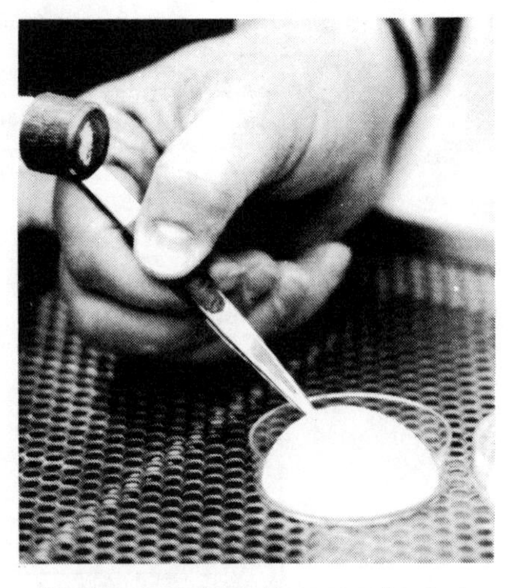

(b) Placement of membrane on pad soaked with culture medium

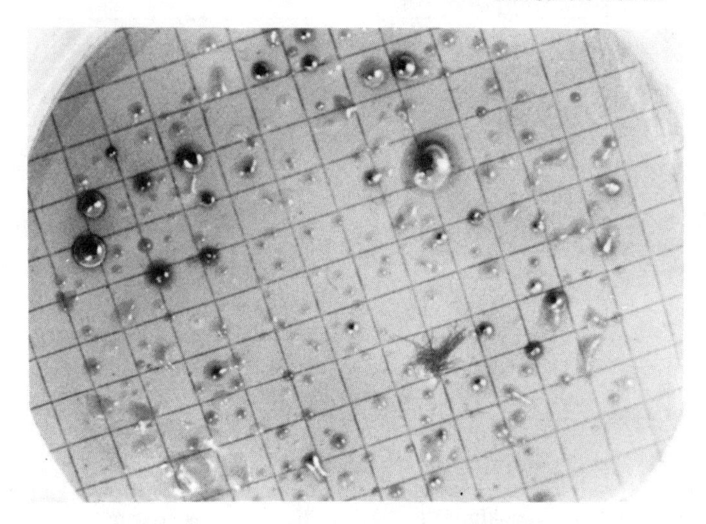

(c) Membrane filter after incubation

Figure 4-3. Steps in the Membrane Filter Technique

Membrane filter technique. The membrane filter method of coliform testing begins by filtering a measured volume of sample (100 mL) under a vacuum through a membrane filter (Figure 4-3). The filter is then placed in a sterile container and incubated in contact with a selective culture media. A coliform bacteria colony will develop at each point on the filter where a VIABLE coliform bacteria was left during filtration. After a 24-hour incubation period, the number of colonies per 100 mL is counted.

A typical coliform colony is pink to dark red with a distinctive green metallic surface sheen. All organisms producing such colonies within 24 hours are considered members of the coliform group. For further confirmation, representative colonies are taken from the filter and run through the first two steps of the multiple-tube technique.

4-2. Standard Plate Count

The standard plate count procedure is the only practical way to estimate the total bacterial population of waters. The test determines the total number of bacteria in a sample that will grow under the influence of a selected media.

Bacteria occur singly, in pairs, in chains, and in clusters. It is known that no single food supply allows all types of bacteria to grow and that the incubation temperature and moisture conditions may not suit every type of bacteria. Since all bacteria will not grow under one set of conditions, test results may reveal fewer bacteria than are actually present. Therefore, a standardized procedure must be used to obtain consistent results.

Significance

Because many microorganisms that develop and are counted in the standard plate count are associated with animals or humans, the count can be considered as a pollution indicator. On the other hand, many bacteria commonly found in water occur naturally and are not related to fecal pollution.

The total number of bacteria in water is not monitored under any current potable water standard. However, plate count tests are sensitive to changes in the raw-water quality and are useful for judging the efficiency of various treatment processes in removing bacteria. For example, if a plate count is higher after filtration than before filtration, there may be a bacterial growth on the filter. The problem would probably not show up during routine coliform analysis. Large increases in the number of organisms may also indicate the entrance of surface drainage during periods of rainfall.

Water leaving a treatment plant may have a low bacterial population, but by the time the water reaches the consumer, the bacterial population could be greatly increased by BACTERIAL AFTERGROWTH—bacteria that have been reduced at the plant but have reproduced in the distribution system. Standard plate count determinations taken as water leaves the treatment plant and at the consumer's faucet will indicate whether this problem exists. High plate counts for samples taken in the distribution system indicate that substantial bacterial growth may be occurring in the distribution system. This occurs as a result of dead ends in the system, inadequate chlorination, or recontamination after chlorination.

Properly treated water should have a standard plate count of less than 500/mL. Higher counts indicate an operational problem that should be investigated.

Test Method

The standard plate count is performed by placing diluted water samples on plate-count agar. The samples are incubated for 48 hours to 72 hours. The bacteria colonies that grow on the agar are then counted using bacteria counting equipment. Detailed procedures are given in the latest edition of *Standard Methods for the Examination of Water and Wastewater.* These procedures must be closely followed in order to provide reliable data for water quality control measurements.

Selected Supplementary Readings

Geldreich, E.E. Is the Total Count Necessary? Proc. Water Quality Technology Conf., Paper No. VII, AWWA, Denver, Colo. (Dec. 1975).

Kincannon, D. F. Microbiology and Surface Water Sources. *OpFlow.* 4:12:3 (Dec. 1978).

McKinney, R. E. *Microbiology for Sanitary Engineers.* McGraw-Hill Book Company (1962).

Rock, R. Procedures and Equipment for the Membrane Filter Technique. *OpFlow.* 5:10:3 (Oct. 1979).

SDWA, IPRs, and MCLs Simplified. *OpFlow.* 3:10:5 (Oct. 1977).

SDWA, IPRs, and MCLs Simplified. *OpFlow.* 3:11:6 (Nov. 1977).

Glossary Terms Introduced in Module 4

(Terms are defined in the Glossary at the back of the book.)

Agar
Bacterial aftergrowth
Coliform (total coliform)
Completed test
Confirmed test
Incubated
Indicator organisms
Membrane filter method
Microbiological
Multiple-tube fermentation method
Negative sample
Pathogens
Positive sample
Presumptive test
Standard plate count
Total coliform test
Viable

Review Questions

(Answers to Review Questions are given at the back of the book.)

1. Define pathogen.
2. Why are coliform bacteria considered a good indicator of pathogens?
3. What indicates a positive response in a fermentation tube?
4. When is the confirmed test run in testing drinking water supplies?
5. Why is the detection of only one coliform colony not an indication of unsafe water?
6. What does the standard plate count indicate and how are the results used?

Study Problems and Exercises

1. You have noticed a consistent variation in the bacteriological sample results from 0–15/100 mL for the last three months. After analyzing the results it is evident that the positive samples were collected from four different points near the outer perimeter of your distribution system. What should you do?

2. The following results are from standard plate count testing of samples from your distribution system. Calculate the standard plate count for each test. What is the significance of these results? What action, if any, should be taken?

mL of Sample	*Colonies Counted*	*Standard Plate Count*
0.1	350	—
0.1	400	—
1 mL of sample diluted 1 : 100	50	—

Module 5

Physical/Chemical Tests

Water has a wide range of physical and chemical characteristics that affect its quality and treatability. Physical and chemical testing of drinking water is necessary to assure that treated water is safe and palatable and to monitor the efficiency of the various water treatment processes. Testing of raw water is also required to help determine treatment techniques and chemical dosages.

After completing this module you will be familiar with the TESTS for

- Alkalinity
- Calcium carbonate stability
- Chlorine residual and demand
- Coagulant effectiveness (jar test)
- Color
- Dissolved oxygen (DO)
- Fluoride
- Free carbon dioxide
- Hardness
- Iron
- Manganese
- pH
- Taste and odor
- Temperature
- Total dissolved solids (TDS)
- Trihalomethanes
- Turbidity
- Inorganic chemicals in general
- Organic chemicals in general.

The significance, sampling procedures, and general testing methods for each test are discussed. The operator can do most of these tests with basic lab equipment, and the tests can provide valuable operational information. Complete procedures for all tests are given in *Standard Methods for the Examination of Water and Wastewater* and the other references listed in the introduction to this book.

5-1. Alkalinity

The alkalinity of water is a measure of the water's capacity to neutralize an acid. Therefore, it is related to the water's BUFFERING CAPACITY, that is, its capacity to resist a change in pH as acid is added.[1]

Significance

Alkalinity has little known significance in regards to human health. However, highly alkaline waters are unpalatable and may force consumers to seek other water sources. Also, alkalinity levels affect the efficiency of the coagulation process. This is especially true when aluminum sulfate, or alum, is used as a coagulant. Alum, an acid salt, reacts with the alkalinity in the water supply. If alkalinity is insufficient, coagulation will be incomplete; soluble alum will remain in the water, and the pH can drop to low levels, which can cause corrosion or other problems. If there is not sufficient alkalinity, it may be necessary to add soda ash or lime. Alkalinity must be determined in order to calculate the lime and soda ash needed for effective water softening. It must also be known when determining if the treated water is corrosive. When calculating the Langelier Index (discussed later in this module) alkalinity must also be known.

Sampling

Samples should be collected from raw and finished water. The raw-water sample indicates whether sufficient alkalinity is present for efficient coagulation and flocculation. The finished-water sample indicates whether the water is corrosive or not. At least 100 mL of sample should be collected in either glass or plastic bottles. Samples may be stored at 39° F (4° C) for no more than 24 hours before analysis. No preservatives should be added.

Methods of Determination

Alkalinity determination consists of titrating the sample to an end point determined with a pH meter. Mixed bromcresol green-methyl red indicator solution (METHYL ORANGE) may be used to determine the end point if a pH meter is not available. The sample must not be filtered, diluted, or altered in any way before the test is performed. The analysis determines total alkalinity, reported as "mg/L as $CaCO_3$."

5-2. Calcium Carbonate Stability

The principal scale-forming substance in water is calcium carbonate. Water is considered stable when it will neither dissolve nor deposit calcium carbonate. This is referred to as the calcium carbonate stability, or equilibrium, point. The reactions and behavior of calcium carbonate and calcium bicarbonate are, therefore, important in water supplies. The actual amount of calcium carbonate that will remain in solution in water depends on several factors: calcium content, alkalinity, pH, temperature, and total dissolved solids.

[1] *Basic Science Concepts and Applications*, Chemistry Section, Acids, Bases, and Salts (Alkalinity).

Significance

Scale formation can cause serious problems in water distribution mains and household plumbing systems by restricting flow, plugging valves, and fouling hot-water heaters and boilers. Corrosion can cause premature pipe or equipment failure. Public health and aesthetic problems can also result since pipe materials such as lead, cadmium, and iron will dissolve into the water.

Several methods can be used to determine the calcium carbonate stability of water. A popular method is the LANGELIER SATURATION INDEX (LI, also referred to as the Langelier Index). The index is equal to measured pH (of the water) minus the pH_S (saturation). The pH_S is the theoretical pH at which calcium carbonate will neither be dissolved into or precipitated from water. At the pH_S a water is perfectly stable. Therefore if $pH - pH_S = 0$, the water is in equilibrium and will neither dissolve nor deposit calcium carbonate on the pipes. If $pH - pH_S > 0$ (positive value), the water is not in equilibrium and will deposit calcium carbonate on the surface of mains and other fixtures. If $pH - pH_S < 0$ (negative value), the water is not in equilibrium and will dissolve the calcium carbonate it contacts. No coating will be deposited on the distribution pipes. However, if pipes are not protected, they may corrode.

The calcium carbonate stability of water is maintained in the distribution system by adjusting the saturation index of the water to a slightly positive value. Adjustment is usually made by adding lime or soda ash.

Sampling

Finished water at the treatment plant and in the distribution system should be evaluated for calcium carbonate stability. The evaluation should be conducted routinely. Evaluation is particularly important when the water treatment plant unit processes or chemical doses are changed. If the index indicates unfavorable conditions, process adjustments should be made. It is very important to remember that the LI is only an indicator of stability. It is not an exact measure of corrosion or calcium carbonate deposition.

The saturation index is developed from results of alkalinity, pH, temperature, calcium content, and total dissolved residue (total dissolved solids). Sampling methods for each of these parameters are discussed in other sections of this module.

Methods of Determination

If the temperature, total dissolved residue, calcium content, and alkalinity of the water are known, the pH_s can be calculated. The following expression may be used:[2]

$$pH_S = A + B - \log(Ca^{+2}) - \log\ (\text{alkalinity})$$

In the equation, A and B are constants, and calcium and alkalinity values are expressed in terms of mg/L as calcium carbonate equivalents. Tables 5-1, 5-2, and 5-3 are used to determine the values of the constants and logarithms.

[2] *Basic Science Concepts and Applications*, Chemistry Section, Treatment Processes (Scaling and Corrosion Control).

Table 5-1. Constant *A* as Function of Water Temperature

Water Temperature C	*A*
0	2.60
4	2.50
8	2.40
12	2.30
16	2.20
20	2.10

Table 5-2. Constant *B* as Function of Total Dissolved Solids

Total Dissolved Solids mg/L	*B*
0	9.70
100	9.77
200	9.83
400	9.86
800	9.89
1000	9.90

The actual pH of the water is measured directly with a pH meter, and the LI is calculated using the formula LI = pH − pH_S. For example, for a water having:

Ca^{+2} = 300 mg/L as $CaCO_3$
Alkalinity = 200 mg/L as $CaCO_3$
Temperature = 16° C
Dissolved Residue = 600 mg/L
pH = 8.7

Determine the saturation index, LI

$pH_S = A + B - \log(Ca^{+2}) - \log\ (\text{alkalinity})$
$pH_S = 2.20 + 9.88 - 2.48 - 2.30$
$pH_S = 7.3$
$LI = 8.77 - 7.3 = 1.4$

An LI of 1.4 indicates that this water is scale-forming.

5-3. Chlorine Residual and Demand

The primary purpose for chlorinating drinking water is to prevent the spread of waterborne disease. Also, chlorine reacts with iron, manganese, sulfide, and taste- and odor-producing agents to help improve finished-water quality. Chlorine is frequently added as water enters the water treatment plant (prechlorination) and again just before it leaves the plant (postchlorination).

Table 5-3. Logarithms of Calcium Ion and Alkalinity Concentrations

Ca^{+2} or Alkalinity mg/L as $CaCo_3$	*Log*	*Ca^{+2} or Alkalinity* mg/L as $CaCo_3$	*Log*
10	1.00	200	2.30
20	1.30	300	2.48
30	1.48	400	2.60
40	1.60	500	2.70
50	1.70	600	2.78
60	1.78	700	2.84
70	1.84	800	2.90
80	1.90	900	2.95
100	2.00	1000	3.00

Tests of chlorine levels in the plant and throughout the distribution system are necessary to determine chlorine dosage levels and monitor water quality.

Chlorination kills bacteria and algae that grow on basin walls and filter media. It also oxidizes substances such as iron and manganese, causing them to PRECIPITATE. Finally, it controls some tastes and odors caused by certain organics.

Postchlorination primarily kills bacteria and provides excess chlorine for continued disinfection in the distribution system. Most substances causing chlorine demand, such as organics and iron, should have been removed in the treatment process, so most of the chlorine will remain available to contact and kill bacteria.

Significance

Disinfection destroys pathogenic (disease-causing) organisms in water. Destruction of harmful bacteria by chlorine is directly related to contact time and concentration of chlorine. High chlorine doses with short contact periods will provide essentially the same results as low doses with long contact periods. Chlorination also oxidizes taste- and odor-causing substances such as iron, manganese, and organic compounds, making their removal from the water possible.

Successful chlorination requires that enough chlorine be added to complete the disinfection or oxidation process. However, chlorine must not be added in amounts that are wasteful and create unnecessarily high operational costs. Determining effective and efficient chlorine dosage levels is the responsibility of the plant operator. Chlorine demand tests indicate whether enough chlorine was added to achieve disinfection and oxidation with enough free chlorine remaining to maintain a chlorine residual in the distribution system.

Chlorine residual. There are two types of chlorine residual: combined residual and free available residual. Figure 5-1, explained in the next two paragraphs, indicates what these two residuals are and how they form.

Assume that a water supply has some natural iron, manganese, organics, and

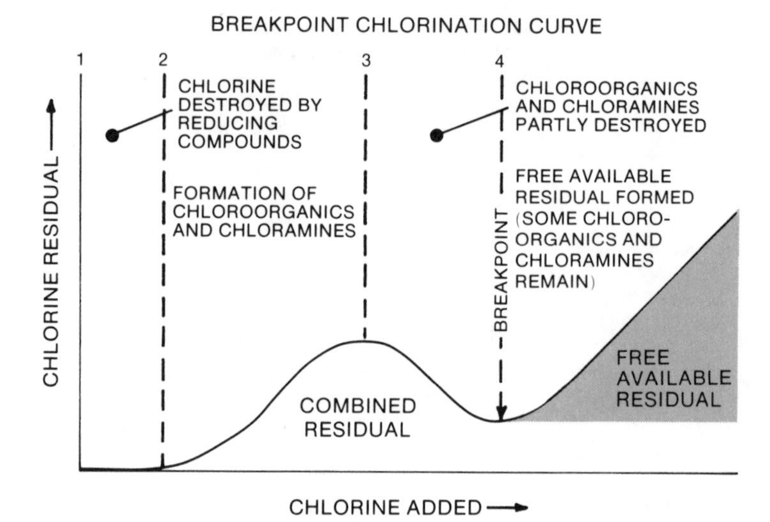

Figure 5-1. Formation of Combined Chlorine Residual and Free Available Chlorine Residual

ammonia. A small amount of chlorine is added—for example, 1 mg/L; it reacts with the iron and manganese only (from point 1 to point 2 in the figure). The chlorine oxidizes the iron and manganese and, in the process, is used up—no residual forms and no disinfection occurs. If the initial chlorine dosage is higher—for example, 2.5 mg/L—the reaction will go to point 3. From point 2 to point 3 the chlorine reacts with the organics and the ammonia, forming chloroorganics and chloramines. These two products are called COMBINED CHLORINE RESIDUAL. This is a chlorine residual that, because it has combined with other chemicals in the water, has lost some of its disinfecting strength. A combined chlorine residual has relatively poor disinfecting power and may cause tastes and odors characteristic of a swimming pool.

As the chlorine dosage is increased further (point 3 to point 4), the chloramines and some of the chloroorganics are destroyed. This reduces the combined chlorine residual until, at point 4, the combined residual reaches its lowest point. Point 4 is called the BREAKPOINT. It marks the point at which the chlorine residuals change from combined to free available. As the initial chlorine dosage is increased still further (beyond 4 mg/L, in this example) FREE AVAILABLE CHLORINE RESIDUAL is formed—"free" in the sense that it has not reacted with anything and "available" in the sense that it can and will react if needed. In terms of disinfecting power, free available chlorine residual is 25 times as powerful as combined residual, and it will not produce the characteristic swimming-pool odor that combined residuals do. Because free available chlorine residual forms only after the breakpoint, the process is called BREAKPOINT CHLORINATION.

The free available chlorine residual at the consumer's tap should be at least 0.2 mg/L. This level helps ensure that the water is free from harmful bacteria. However, higher levels may be necessary to control special problems such as iron bacteria. If a free chlorine residual cannot be maintained in a distribution system, several possible problems are indicated. Dead ends, biological growths, breaks, or cross connections all cause dissipation of free chlorine residuals.

Chlorine demand. Test results for chlorine residual can be combined with operating data regarding the amount of chlorine added at the plant to yield information on CHLORINE DEMAND. Chlorine demand is a measure of how much chlorine must be added to the water to achieve breakpoint chlorination or whatever free chlorine residual is desired.[3]

The most significant reason for conducting analysis of a water supply's chlorine demand is to determine the proper dosage. However, changes in chlorine demand can also indicate water-quality changes. For example, if a water supply suddenly requires more chlorine in order to maintain a residual (that is, the water exhibits a higher chlorine demand) then the chlorine is oxidizing some contaminants that previously did not exist in the water supply. When this occurs, two steps are necessary. First, the chlorine dose must be increased to meet the higher demand. Second, the reason for the increased demand should be investigated. This sudden increase in chlorine demand frequently occurs because of seasonal water-quality changes in surface water. Chlorine demand in ground water should not change substantially because the quality of ground water is usually very stable.

Sampling

Chlorine residual sampling is done at the consumer's faucet and at the plant. Sampling at the consumer's faucet is done to determine whether consumers are receiving water that is safe to drink. These analyses, if made at the same time and location as bacteria samples, may be correlated with coliform results.

A 100-mL sample is sufficient for measuring free available chlorine residual using either the DPD (n,n-diethyl-p-phenylenediamine) or amperometric method. Most DPD color comparator kits require a sample volume less than 100 mL. A field test kit vial, which requires about 10 mL of sample, can be used. Analysis should be completed as soon as possible after collection. Preservation is not recommended since chlorine is unstable in water and residual chlorine will continue to diminish with time. Agitation or aeration of the sample should be avoided because this can cause reduction of the sample's chlorine concentration. Chlorine in samples exposed to sunlight will also be destroyed and subsequent analysis will be erroneously low. Never use the same sample bottles for chlorine and coliform analysis. Bottles used for coliform analysis contain a chemical (sodium thiosulfate) that destroys chlorine residuals.

In-plant sampling of chlorine residual determines whether sufficient chlorine has been added to the water before it leaves the treatment plant. This is the only way to be sure that finished water leaving the plant contains the desired chlorine residual. Obtaining representative samples is the most critical part of in-plant chlorine sampling.

Often the sample must be collected at a point near the location of chlorine

[3] *Basic Science Concepts and Applications*, Chemistry Section, Dosage Problems (Chlorine Dosage Demand/Residual Calculations).

addition. These samples will probably show an unrealistically high chlorine residual. To obtain data that approximates the actual chlorine residual in the particular basin, the sample should be held for a time period equal to the basin detention time, or in any case at least 10 min.

Methods of Determination

The DPD color comparator test kit is the simplest and quickest way to test for residual chlorine. The test takes approximately 5 min to complete. General procedures for DPD testing are given in Appendix C. The orthotolidine (OT) method should not be used since it is not as accurate as the DPD method, particularly for free chlorine residual. In addition, the chlorine demand should be determined routinely to determine proper dosage. The chlorine demand can be determined by treating a series of water samples with known but varying chlorine dosages. After a desired contact time, the chlorine residual of each sample is determined. This will indicate which dosage satisfied the demand and provided the desired residual.

Another technique, used primarily in laboratories because of its accuracy, is amperometric titration. The method is unaffected by sample color or turbidity that interfere with colorimetric determinations. However, performance of amperometric titration requires great skill and care.

5-4. Coagulant Effectiveness (Jar Test)

One of the objectives of a water treatment plant is to produce clear, colorless water. Cloudy or turbid water attracts complaints from consumers and poses a variety of potential health hazards. Surface waters, in particular, contain silt and other suspended matter that must be removed during the treatment process. Removal of these particles is accomplished by coagulation, flocculation, and sedimentation, followed by filtration. For most effective and efficient filter performance, the turbidity of water entering the filters should be 10 NTU (NEPHELOMETRIC TURBIDITY UNITS) or less. The jar test is used to determine the proper coagulants and dosages needed to achieve this level.

Significance

Coagulation/flocculation involves addition of chemical coagulants such as aluminum sulfate, ferric chloride, or polyelectrolytes. The most efficient plant operation occurs when the lowest turbidity is obtained at the lowest coagulant dosage. The jar test provides the information necessary to accomplish this goal. By approximating the conditions that exist in the water treatment plant, the test allows operators to select optimal chemical doses in the laboratory, rather than by trial and error in the plant. The test can also be used to check the adequacy of flash mixing or flocculation mixing in the plant. Test results may indicate the need to improve or modify flash mixers and flocculation basins to obtain more efficient operation.

Figure 5-2 illustrates typical jar-test results from a water supply in which aluminum sulfate is used as a coagulant. The data show that the benefits of the

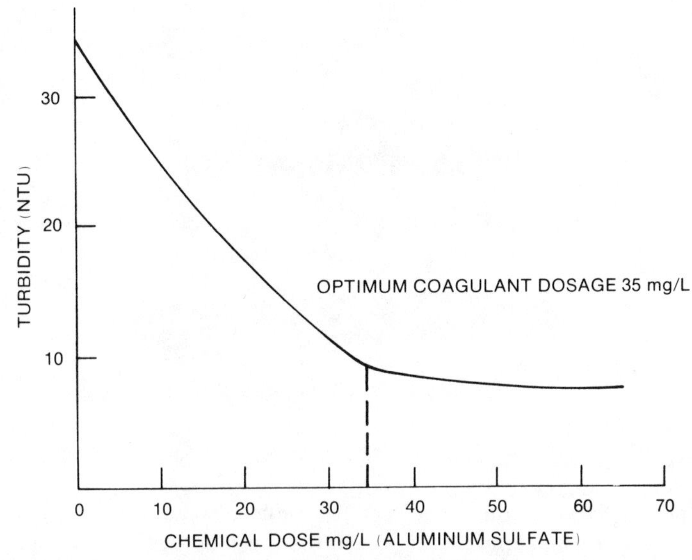

Figure 5-2. Typical Jar Test Results

coagulant decrease with doses beyond 35 mg/L. In other words, beyond a dose of 35 mg/L, it takes a very large increase in chemical dosage to produce a small increase in turbidity removal. Therefore, 35 mg/L should be considered as the optimal dose for the water tested.

When evaluating several coagulants, the goal is to identify the one coagulant or combination of a coagulant and coagulant aid that will produce a low turbidity using the least expensive dosage of chemicals. Chemical prices must also be evaluated since a low dosage of an expensive chemical can be more costly than a high dosage of an inexpensive chemical.

Sampling

Water samples for jar test analysis should be collected immediately before the point or points of coagulant chemical application. Larger samples (a minimum of 6 L for most tests) are required for jar tests than for most chemical analyses.

Test results are affected by temperature changes of the sample or by the length of time the sample is held. Consequently, tests should be run as soon as possible after collection. The test should be conducted whenever there is a significant change in water quality or other conditions that may require a change in coagulant dosage.

Methods of Determination

There is no standard procedure for conducting a jar test, nor is there standard test equipment that must be used. Best results are usually obtained with a commercially manufactured jar test unit. Whatever equipment is used, the tests for a particular plant should always be conducted using a uniform, well-defined

Figure 5-3. Jar Test Apparatus

procedure so as to allow comparison of the most recent results with the results of past analyses.

A jar test apparatus consists of a stirring machine with three to six paddles (Figure 5-3). The unit should have a variable speed control to allow mixing at 0 to 100 rpm. After mixing and settling, the floc in each beaker is observed. This visual observation gives a rough idea of what dosage provides the best coagulation. Visual observation does not, however, give precise results that can be compared with previous tests. Samples may appear clear even though small, hard-to-filter particles still remain in suspension. Therefore, turbidity measurements should be made with a NEPHELOMETRIC TURBIDIMETER in order to ensure accurate evaluation of jar test results. Sampling for turbidity analysis should be performed in a uniform manner. The tip of a volumetric pipet is placed approximately 1 in. (25 mm) below the surface of the sample and approximately 100 mL of sample is slowly withdrawn for analysis.

A typical procedure for jar test determinations, taken from AWWA Manual M12, *Simplified Procedures for Water Examination*, is given in Appendix D of this book.

5-5. Color

Color in water is caused by minerals, aquatic life, or organic matter from soils and vegetation. Color can also be caused by industrial or municipal contamination. Color is usually only a problem with surface waters, but some ground waters containing iron or manganese can also have significant color levels.

Color in water is classified as either true color or apparent color. True color is due to colloidal organic compounds in the water. Apparent color is caused by colored suspended matter such as clay or iron precipitates. In water treatment applications, true color is the most difficult to remove.

Color is measured by comparing the color of a sample with the color of a standard chemical solution. The unit of measurement is the COLOR UNIT, or CU.

Usually a water with a true color less than 15 CU passes unnoticed while a true color of 100 CU has the appearance of tea.

Significance

Color definitely affects consumer acceptance of drinking water. Consumers will reject colored water and may change to another source of water even if it is less safe. Color is also an indication of high levels of organic compounds, which may produce trihalomethanes upon contact with chlorine. Color in drinking water should be removed to produce a pleasing, acceptable appearance.

Sampling

Color determinations should be routinely completed on raw water, finished water, and on water from the distribution system. Sampling in surface-water systems is especially important. Color data from raw- and finished-water sample points indicate treatment plant efficiency.

Color samples taken at the consumer's tap help in detecting organic growths or corrosion in the distribution system. At least 50 mL of sample should be collected in a clean glass or plastic bottle. The sample should be cooled to 39° F (4° C), and the analysis completed within 24 hours.

Method of Determination

The PLATINUM-COBALT METHOD is the preferred method for color analysis. It is useful for measuring color derived from naturally occurring materials but it is not applicable to color measurement on waters containing highly colored industrial wastes.

5-6. Dissolved Oxygen (DO)

Dissolved oxygen (DO) in water is not considered a contaminant. An excess or lack of dissolved oxygen does, however, help create unfavorable conditions. Generally, a lack of dissolved oxygen in natural waters creates the most problems, specifically, an increase in tastes and odors as a result of ANAEROBIC decomposition.

The amount of dissolved oxygen in water is a function of the water's temperature (colder water contains more dissolved oxygen) and salinity (more saline water contains less dissolved oxygen). Natural waters are seldom in EQUILIBRIUM (exactly saturated with dissolved oxygen). Temperature changes as well as chemical and biological activities all use or release oxygen, causing the amount of dissolved oxygen in water to change continually.

Significance

Dissolved oxygen in municipal water supplies is generally not a problem. It has no adverse health effects and actually increases the water's palatability. Most consumers prefer water that has a dissolved oxygen content near saturation. However, a concentration this high is detrimental to metal pipes because oxygen helps accelerate corrosion.

Oxygen is important as an OXIDANT in water plant operation. Its primary value is to OXIDIZE iron and manganese into forms that will precipitate out of the water. It also removes excess carbon dioxide. Provided there is long enough contact time, it will also help degrade some organic compounds that cause taste and odor problems. Usually some form of aeration is used to ensure that enough dissolved oxygen is present in the water for effective oxidation. In addition, operators can use dissolved oxygen data from their raw-water storage reservoirs to get an indication of the general quality of the water. Based on this data, operators may be able to make treatment changes or alter the manner in which the reservoir releases are made in order to prevent taste, odor, and other problems.

Sampling

Analyses should be conducted routinely on raw-water samples, particularly if storage reservoirs are being used. Treated-water samples should also be analyzed routinely if aeration is used as a treatment process; otherwise the tests can be conducted on a weekly basis for general quality data.

Dissolved oxygen should be determined on site if the electrode method is used. If the Winkler test is used, the sample must be collected in a glass BOD bottle and "fixed" on site. The sample should be stored in the dark at the temperature of the collection water or water-sealed and kept at a temperature of 50 to 68° F (10 to 20° C).

Methods of Determination

The ELECTRODE METHOD and the MODIFIED WINKLER METHOD (also called the IODOMETRIC METHOD) are preferred for dissolved oxygen measurements. Because the electrode method is not as sensitive to interferences as the modified Winkler test, it provides an excellent method of dissolved oxygen analysis in polluted waters, highly colored waters, and strong waste effluents. Drinking waters and supply reservoirs have few interferences that cause problems with the modified Winkler (azide modification) procedure. Therefore, the azide modification and electrode methods are generally acceptable for dissolved oxygen analysis of drinking water.

5-7. Fluoride

Fluoride is found naturally in many waters. It is also added in many water systems to reduce tooth decay.

Significance

Research has demonstrated that drinking water containing a proper amount of fluoride during the years of teeth formation (from birth to 12 to 15 years) results in an average of 65 percent reduction in tooth decay.

Optimum fluoride concentrations in drinking water to reduce tooth decay vary with climate. Because more water is consumed in warmer climates, fluoride concentrations should be lower in these areas. Excessive fluoride concentrations can cause teeth to become stained or MOTTLED. This is generally only a problem where natural fluoride concentrations exceed 2.4 mg/L. However, close control of fluoride concentrations is necessary to assure the maximum benefit of fluoridation with an adequate margin of safety. A drop of only 0.3 mg/L below the optimum concentration can drastically reduce the dental benefits of fluoride. Specific recommendations from the state should be obtained concerning recommended concentrations for a given water supply, particularly if fluoride is being added to the water.

Sampling

Fluoride samples should be taken on raw and finished waters. Raw-water samples are necessary because the total amount of fluoride reaching the consumer is equal to the raw-water fluoride concentration plus that added at the plant. The amount of fluoride to be added to the raw water is calculated by subtracting the raw-water concentration from the desired treated-water concentration.

Finished-water samples are tested to ensure that the fluoride feeders are operating correctly and the final fluoride concentration is at the desired level. Daily samples should be tested.

Samples collected for fluoride analysis may be held for seven days before analysis. They should be stored in a refrigerator at 39° F (4° C) with no preservatives added. Glass or plastic bottles can be used for collection and storage.

Methods of Determination

There are two commonly used methods for fluoride analysis: the SPADNS METHOD [sodium, 2-(parasulfophenylazo)-1,8-dihydroxy-3, 6-naphthalene disulfonate] and the ELECTRODE METHOD. The SPADNS method requires a time-consuming distillation step to eliminate interferences and is not recommended if an electrode is available.

The electrode method requires a selective ion fluoride electrode connected to a pH meter with a millivolt scale or to a meter having a direct concentration scale for fluoride.

5-8. Free Carbon Dioxide

Carbon dioxide, a colorless, odorless, noncombustible gas, is found in all natural waters. Carbon dioxide in surface waters can originate from the atmosphere but most comes from biological oxidation of organic matter. Biological oxidation is also the primary source of carbon dioxide in ground waters.

Significance

No negative public health effects have been found from consuming excess carbon dioxide. In fact, carbon dioxide is present in carbonated beverages in concentrations far greater than those found in natural waters. However, carbon dioxide in water can cause corrosion problems. In addition, carbon dioxide values must be known to calculate proper lime dosages when softening water. If RECARBONATION is used following lime softening, carbon dioxide values should also be determined in order to better control the process.

Sampling

Carbon dioxide analyses should be run on raw and finished water. Special precautions must be taken during collection and handling of the sample if using the titrimetric method. Exposure to the air must be kept to a minimum. Field determination of free carbon dioxide immediately after sampling is advisable. If field determination is impossible, keep the sample cool and complete the analysis as soon as possible.

Samples may be collected in glass or plastic bottles. At least 100 mL of sample should be collected. The bottle should be filled to the top with no air space and no preservatives should be added.

Methods of Determination

The amount of carbon dioxide in water may be determined by using the NOMOGRAPHIC METHOD or TITRIMETRIC METHOD. In order to use the nomographic method, pH, bicarbonate alkalinity, temperature, and total filterable residue (total dissolved solids) must be known. Results are most accurate when the pH and alkalinity are analyzed immediately after sample collection. The titration method may be performed POTENTIOMETRICALLY or with PHENOLPHTHALEIN INDICATOR.

5-9. Hardness

Hardness is a measure of the concentration of calcium and magnesium salts in water.[4] They are generally present as bicarbonate salts. Water hardness is derived largely from contact with soil and rock formations. Hard waters usually occur where topsoil is thick and limestone formations are present. Soft waters occur where the topsoil is thin and limestone formations are sparse or absent.

Significance

Hard and soft waters are both satisfactory for human consumption. However, consumers may object to hard water because of scaling problems it causes in household plumbing fixtures and on cooking utensils. Hardness is also a problem for industrial and commercial users because of scale buildup on boilers and other equipment.

Water most satisfactory for household use contains about 75 to 100 mg/L as $CaCO_3$. Waters with a hardness of 300 mg/L as $CaCO_3$ are generally considered too hard. A recent trend in water plant softening has been to partially soften water to 75 to 150 mg/L as $CaCO_3$, which reduces chemical costs over complete

[4] *Basic Science Concepts and Applications*, Chemistry Section, Solutions (Hardness).

softening and provides a water acceptable to the consumer.

Very soft waters, found in some sections of the United States, have hardness concentrations of 30 mg/L as $CaCO_3$ or less. These waters are generally corrosive and are sometimes treated to increase hardness.

Sampling

The number and location of hardness sampling points in a system depends on whether a plant practices water softening. If a plant softens water, hardness analysis on finished water should be conducted daily to determine whether the desired degree of softening has been achieved. Analyses should be conducted immediately after filtration and before the water enters a clearwell.

Hardness determinations should also be performed on raw waters whenever weather conditions, such as spring rains, affect the supply. This type of sampling would reveal any variation in the hardness of the raw water and provide advance information for chemical dosage changes that may be necessary for softening. Even if softening is not practiced, hardness determinations should still be made periodically as a general water-quality measure.

At least 100 mL of sample should be collected in either glass or plastic bottles. Samples may be stored for no longer than seven days before analysis. They should be cooled to 39° F (4° C) and acidified with 5 mL/L (0.5 mL/100 mL) of nitric acid, unless the samples are going to be analyzed at once.

Methods of Determination

The EDTA (ethylenediaminetetraacetic acid) titrimetric method is the preferred method of analysis. It consists of SEQUESTERING (tying up) the calcium and magnesium ions by titrating with an EDTA solution. The sample is titrated in the presence of an INDICATOR. The initial solution is red and changes to blue when all the ions have been sequestered.

5-10. Iron

Iron occurs naturally in rocks and soils and is one of the most abundant of all elements. It exists in two forms: "ferrous" (Fe^{+2}) and "ferric" (Fe^{+3}) iron. Ferrous iron is found in well waters or in waters without much dissolved oxygen. Under ANAEROBIC conditions waters can have significant iron concentrations.

Iron in solution in water is derived naturally from soils and rocks. It may also result from the corrosive action of water on unprotected iron or steel mains, steel well casings, and pumps. Surface waters may contain appreciable amounts of iron originating from industrial wastes or from acid runoff from mining operations.

Significance

There are no harmful health effects from drinking water containing iron. Water-quality limits on allowable concentrations of iron in water supplies are based on aesthetic and taste problems rather than health concerns. Iron concentrations above 0.3 mg/L can cause "red water" and staining of plumbing fixtures. Concentrations at or above this level in finished water indicate that steps should be taken to provide iron removal.

Iron also provides a nutrient source for some bacteria that grow in distribution systems and wells. IRON BACTERIA, such as *Gallionella*, cause red water, tastes and odors, clogged pipes, and pump failure.

Whenever tests show increased iron concentrations from the plant to the consumer's tap, corrosion and/or iron bacteria may be present, and corrective action should be taken. If the water is corrosive, pH adjustment might first be considered. If the problem is caused by bacteria, flushing of the mains, shock chlorination, or increased chlorination may prove effective.

Sampling

Samples should be collected from raw and finished water. The samples should be collected in glass or plastic bottles and may be stored up to six months before analysis. At least 100 mL of sample should be collected. Samples should be preserved with concentrated nitric acid. Approximately 5 mL/L (0.5 mL/100 mL) should be added to lower the pH to less than 2.

Methods of Determination

Iron concentration may be determined by the PHENANTHROLINE METHOD or the ATOMIC ABSORPTION SPECTROPHOTOMETRIC METHOD (described under 5-18, Inorganic Chemicals in General). The phenanthroline method is simple and reliable. It is a colorimetric test and can be run with a spectrophotometer, filter photometer, or Nessler tubes. The atomic absorption method, used by most laboratories, is very accurate and particularly advantageous when large numbers of samples must be tested.

5-11. Manganese

Manganese creates problems in a water supply similar to those created by iron. Manganese, a metal, occurs naturally in ores but not in a pure state. It exists in soils primarily as manganese dioxide. Manganese is found in both the manganous, divalent form (Mn^{+2}) and in the quadrivalent form (Mn^{+4}). Manganese is much less abundant in nature than iron; therefore, it is less often present in water supplies and exists at lower concentrations. It is also more difficult to oxidize or cause to precipitate. Since manganous solutions are more STABLE than ferrous solutions, manganese removal procedures are more complicated.

The most common forms of manganese—oxides, carbonates, and hydroxides—are only slightly soluble. Consequently, manganese concentrations in surface water seldom exceed 1.0 mg/L. In ground waters subject to anaerobic or reducing conditions, manganese concentrations, like iron concentrations, can become very high.

Significance

Manganese has no harmful effects when consumed by humans. Water-quality limits on allowable concentrations of manganese have been based on aesthetics rather than health concerns. Manganese does not usually discolor the water, but it stains clothes and bathroom fixtures black. Staining problems begin at 0.05 mg/L, a much lower concentration than for iron.

Raw- and finished-water analyses will indicate whether manganese removal is necessary or whether the desired manganese removal has been achieved in the treatment plant. Increases in manganese concentration in the distribution system are not generally experienced. Rapid flow changes in the distribution system may result in some deposits breaking loose and entering the consumer's home. This problem is best controlled by flushing the lines in areas where the problem occurs.

Sampling

Samples should be collected from raw and finished water. The samples should be collected in glass or plastic bottles and may be stored up to six months before analysis. At least 100 mL of sample should be collected. Samples should be preserved with concentrated nitric acid. Approximately 5 mL/L (0.5 mL/100 mL) should be added to lower the pH to less than 2.

Methods of Determination

The ATOMIC ABSORPTION SPECTROPHOTOMETRIC METHOD (described under 5-18, Inorganic Chemicals in General) is the preferred method of determination.

5-12. pH

The pH is a measure of water's acidity or alkalinity.[5] A scale of 0 to 14 is used for measurement, with 0 being extremely acidic and 14 being extremely alkaline. The midpoint, 7, is neutral. Intensity increases 10 times for each drop or rise of one unit. For example, a pH of 5.0 indicates 10 times more acidity than a pH of 6.0.

Significance

The efficiency of chlorination, coagulation, softening, and corrosion control processes depends on the proper pH. In addition, monitoring pH throughout the treatment process may help to locate a failure in a particular process. For example, where lime softening is practiced, pH should increase significantly after lime has been added. If the pH does not increase, insufficient or faulty lime additions could be indicated. Monitoring can also indicate changes in raw water. For example, introduction of an industrial waste discharge may lower the pH.

[5] *Basic Science Concepts and Applications*, Chemistry Section, Acids, Bases, and Salts (pH).

Sampling

Samples should be taken from raw water, finished water, and water at various treatment stages. Such extensive monitoring helps in controlling the efficiency of these processes and aids in determining whether artificial alteration of pH is necessary to obtain a suitable finished water.

Samples may be collected in glass or plastic containers. At least 25 mL of sample should be taken. Contamination of the samples must be avoided because small quantities of contaminants can drastically change results. In addition, samples should be collected without agitation because a loss of carbon dioxide caused by aeration can increase the pH.

The test should be run as soon as possible after collection. Samples taken in the field should be analyzed there with a portable meter, if one is available. If the sample must be transported to the laboratory, it should be cooled to 39° F (4° C) with no preservatives added. The test should be performed no longer than six hours after collection.

Methods of Determination

The two commonly used methods for determining pH are (1) COLORIMETRIC METHODS, which use color indicators or discs, and (2) the PROBE METHOD.

Colorimetric methods are less expensive than the probe method; however, they are subject to interference due to color, turbidity, and dissolved solids found in the water. Also, the indicator solutions used in colorimetric determinations have several disadvantages: the indicator may deteriorate with time or it may slightly alter the pH of the water being tested. Consequently, the colorimetric method is not recommended for use in water treatment.

The probe, or electrometric, method is the recommended procedure. It is quick, easy, and convenient, and it is unaffected by outside interferences. The method uses an electrode and a meter CALIBRATED with a solution of known pH. Proper calibration is essential for accurate measurements. The electrode is placed in the water sample and the pH is read directly from the meter.

5-13. Taste and Odor

Tastes and odors in water are difficult to measure. They are caused by a variety of substances, including organic matter, dissolved gases, and industrial wastes. Odors in water supplies are frequently caused by algae or decaying organic matter. Intensity and offensiveness of odors vary with the type of organic matter. Odors are classified as aromatic, fishy, grassy, musty, septic, or medicinal. Industrial wastes, such as phenolic or oil waste, are also responsible for some odors in surface waters.

The human sense of smell is much more sensitive than the sense of taste, and odor tests are commonly run in water treatment plants. The taste test, which classifies tastes as sweet, sour, bitter, and salty, can only be run on water known to be safe for drinking; this limits its usefulness.

Significance

The odor test can be used to evaluate how well a water treatment plant removes taste- and odor-causing organic materials. It can also be used to trace contamination sources that may be entering a plant. For example, an odor-causing industrial-waste discharge might be occurring upstream from a water treatment plant. Samples are collected at intervals upstream until the problem-causing area has been reached. The odors should become stronger closer to the discharge and should not be evident in samples collected upstream from the discharge. This technique is time consuming; however, it can be conducted by water plant personnel, and extensive laboratory facilities are not necessary.

The odor test can also be used to detect problems in the distribution system. For example, odors will occur in dead-end lines with significant bacteriological buildups. A definite chlorine odor can indicate the loss of free residual due to stagnation, slime buildup, and ANAEROBIC conditions.

The THRESHOLD ODOR NUMBER (TON) is designed to help monitor all types of odors, independent of source. The TON cannot, however, be compared to the concentration of the odor-producing substance because some substances produce strong odors at low concentrations. For example, some chemical wastes, such as phenol, in chlorinated water have been detected by the threshold odor test at a 0.001 mg/L concentration. Other odor-producing substances, such as detergents, may not be detected until the concentration is as high as 2.5 mg/L.

An odor with a TON of 3 might be detected by a consumer whose attention was called to it but would not be noticed otherwise. If an odor appears gradually, the consumer will adapt to it and the odor will be noticed less than if it appears suddenly. Finished-water quality with a TON above 5 will begin to draw complaints from consumers. When a TON of 3 or more is detected in a finished-water supply, quick action should be taken to solve the problem.

Sampling

Odor tests should be conducted routinely by plant personnel. Water supplies with seasonal or recurring taste and odor problems should be analyzed regularly, and as problems occur, corrective action should be taken. The test generally takes several hours, and it is usually not possible to conduct more than one or two tests per day.

Water samples should be taken from raw and finished waters. At least 1000 mL of sample should be collected for an odor analysis. Samples should be collected in clean bottles that have not been used for any samples that might leave a taste or odor. Bottles should be washed with detergent and rinsed with distilled water. Glass sample bottles are recommended. Plastic containers should not be used since plastic will give the water taste and odor.

Aeration and mixing of the sample should be kept to a minimum before testing because air will STRIP, or OXIDIZE, odor-producing compounds. An air space should be left at the top of the bottle so that the sample can be thoroughly shaken before testing.

Odor tests should be run as soon as possible after collection. If the sample must be stored, it should be tightly capped and placed in an odor-free refrigerator. The sample should be analyzed no later than 24 hours after collection.

Methods of Determination

Most tastes and odors are extremely complex, and the best way to detect them is with the human sense of smell. A series of sample dilutions are prepared and placed in bottles for observers to test. Each bottle contains 200 mL of liquid consisting of a mixture of sample and odor-free distilled water. The bottles are arranged so that the observers smell the most dilute samples first. Observers do not know the dilution ratios in the bottles. The dilution ratio at which an observer first detects an odor is called the threshold odor number. The TON may be calculated as follows:

$$\frac{V_s + V_d}{V_s} = \text{Threshold Odor Number}$$

Where: V_s = Volume of sample (mL)
V_d = Volume of dilution water (mL).

The lowest obtainable TON is 1. If no odor is detected in a 200 mL sample (a sample containing no odor-free distilled water), the TON is reported as "no odor observed" and an actual number is not assigned. For example, if an odor is first detected in a bottle that has 100 mL of sample dilute to 200 mL with distilled water, the TON is 2. Threshold odor numbers corresponding to various dilutions are shown in Table 5-4.

The threshold odor test is not precise and represents human judgment. The ability to detect odors varies, and panels of five or more persons are recommended to overcome this variability. At least two persons are required to run an odor test. Persons performing odor tests should not have colds or allergies that would affect their sense of smell. They should be nonsmokers since smoking dulls the sense of smell. Plant operators should not make odor

Table 5-4. TON Corresponding to Various Dilutions

Sample Volume (mL) *Diluted to 200 mL*	*TON*
0.8	256
1.6	128
3.1	64
6.3	32
12.5	16
25	8
50	4
100	2
200	1

observations since they work with the water in question and their sense of smell is dulled. All tests should be conducted in an odor-free atmosphere.

5-14. Temperature

Temperature is measured on either of two scales: the fahrenheit (F) scale or the celsius (C) scale. The freezing point of water is 32° F, or 0° C; the boiling point is 212° F, or 100° C.[6]

Significance

Water temperature determines, in part, how efficiently certain unit processes operate in the treatment plant. The rate at which chemicals dissolve and react is slightly dependent on temperature. Cold water requires more chemicals for efficient coagulation and flocculation, whereas warm water requires less chemicals for these processes to be efficient. High water temperatures may indicate a greater chlorine demand due to increased organics, such as algae, in the raw water. Water temperatures also determine the speed at which plant and animal life will grow, with growth being slow in cold water and rapid in warm water.

Sampling

Temperature readings must be taken on site, either directly from the water or from samples immediately after collection. Immediate readings are necessary because water temperature will begin to change once the sample is taken.

Methods of Determination

A thermometer is used for temperature analyses. The thermometer is left in the water long enough to get a constant reading. The measured temperature should be expressed to the nearest degree or less depending on the thermometer's accuracy.

5-15. Total Dissolved Solids (TDS)

Total dissolved solids (TDS), also referred to as total filterable residue in natural waters consists mainly of carbonates, bicarbonates, chloride, sulfate, calcium, magnesium, sodium, and potassium. Dissolved metals, dissolved organics, and other substances account for a small portion of the dissolved residue in water.

Significance

Dissolved residue in drinking water tends to change the water's physical and chemical nature. Distilled or deionized water has a flat taste, whereas water with some dissolved solids is preferred by most consumers. Different salts in solution may interact and cause effects that each salt alone would not cause. The presence of harmful dissolved compounds or ions (such as arsenic and mercury) can be

---

[6] *Basic Science Concepts and Applications*, Mathemetics Section, Conversions (Temperature Conversions).

dangerous in water even where the total solids concentration is relatively low.

It is generally agreed that the TDS concentration of palatable water should not exceed 500 mg/L. Lime softening and ion exchange facilities both significantly reduce the quantity of TDS in finished water.

Many communities in the United States use waters containing 2000 mg/L or more TDS because better water is not available. These waters tend to be unpalatable, may not quench thirst, and can have a laxative effect on new or transient users. No lasting harmful effects have been reported from these waters. Waters containing more than 4000 mg/L TDS are considered unfit for human consumption.

Sampling

Samples should be taken on raw and finished waters. At least 100 mL of sample should be collected in either a glass or plastic bottle. The sample should be preserved by cooling to 39° F (4° C) and should be analyzed within seven days after collection.

Methods of Determination

A GRAVIMETRIC PROCEDURE is the preferred method of determination. A well-mixed sample is filtered through a standard glass fiber filter. A measured quantity of the filtrate is evaporated and dried to constant weight at 356° F (180° C). Results are only an indication of the anticipated palatability of the water; they do not indicate whether the water is safe or what substances it contains.

5-16. Trihalomethanes

Trihalomethanes are compounds formed when chlorine reacts with humic and fulvic acids—natural organic compounds that come from decaying vegetation. Chloroform ($CHCl_3$) is the most common trihalomethane found in drinking water, but bromodichloromethane ($CHBrCl_2$), dibromochloromethane ($CHBr_2Cl$), and bromoform ($CHBr_3$) are frequently found as well (Figure 5-4).

Significance

Chloroform has been shown to cause cancer in laboratory animals; therefore it is considered to have a high potential for causing cancer in humans. Since the other trihalomethanes are similar to chloroform, they are also suspected of causing cancer. Trihalomethanes are found in practically every water system using chlorine, so a very high percentage of the American public is exposed to these compounds in their drinking water.

Sampling

Special glass vials with a volume of at least 25 mL and screw-caps lined with PTFE are used for THM samples. These vials should be obtained from the laboratory that will perform the analysis.

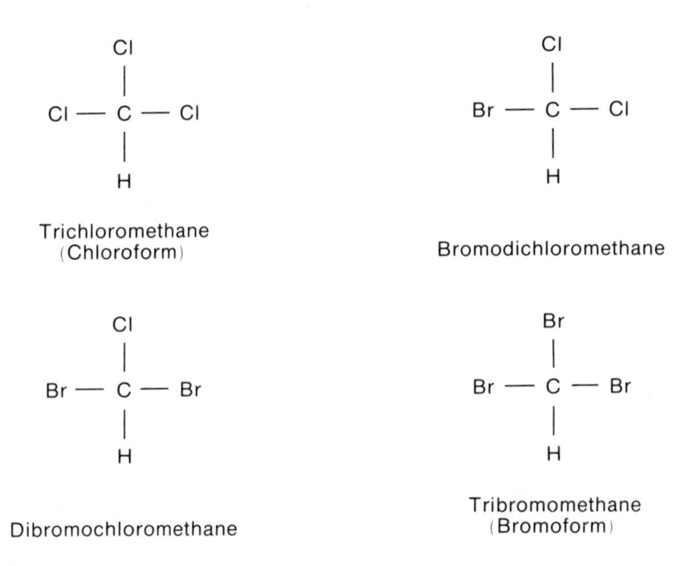

Figure 5-4. Common Trihalomethanes

If the THM concentration at the time of sampling is desired, a chemical reducing agent (sodium thiosulfate or sodium sulfite) must be added to the sample to stop the formation of THM after sample collection.

The sample vial must be filled so that no air bubbles pass through the sample as the vial is filled. To check the sample, invert the vial once the cap is on. The absence of entrapped air indicates a successful seal. If a chemical reducing agent has been added, the vial must not be shaken for about 1 min after it has been capped.

The samples should be analyzed within 14 days. If the samples must be shipped or stored, the laboratory that will perform the analysis should be asked for instructions on shipping and storage.

Methods of Determination

Analysis for trihalomethanes requires the use of a GAS CHROMATOGRAPH by an experienced chemist. Consequently, only major water utilities, state agencies, and commercial laboratories commonly conduct trihalomethane analyses. Smaller water utilities usually send their samples out to be analyzed.

A gas chromatograph separates complex mixtures of chemical compounds from one another by SELECTIVE ADSORPTION and identifies them with a detector. The sample is injected into a heated block where it is vaporized. The vapor is swept through a special column packed with adsorptive material where the chemical components of the vapor are separated and detected. The detector feeds this information into a recorder. The recorder then translates the information into a series of peaks; each peak corresponds to a compound. The

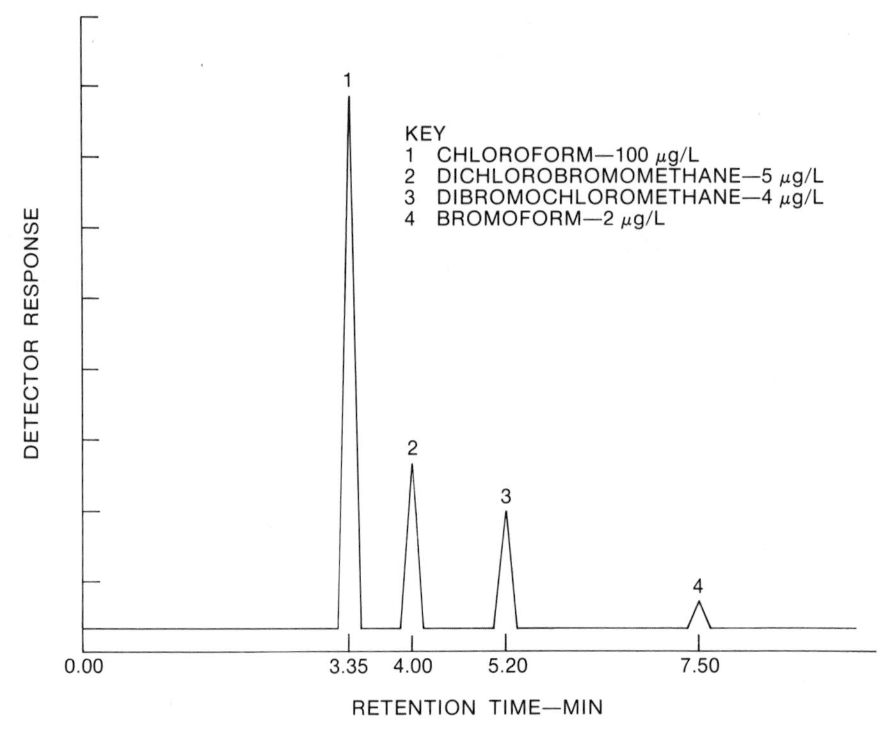

Figure 5-5. Sample Readout from a Gas Chromatograph

area under each peak corresponds to the concentration of that particular compound. A sample readout from a gas chromatograph is shown in Figure 5-5.

5-17. Turbidity

Turbidity is caused by suspended particles in water. These particles scatter or reflect light rays directed at the water, making the water appear cloudy. Waters causing very little light scattering produce low turbidity measurements, whereas those causing a great deal of light scattering produce high turbidity measurements. The suspended particles causing turbidity include organic and inorganic matter and plankton.

Turbidity should not be confused with suspended solids. Turbidity expresses how much light is scattered by the sample, whereas suspended solids expresses the weight of suspended material in the sample. Since there is no direct relationship between suspended solids and turbidity, exact comparisons between the two are difficult to make.

Significance

Turbidity is significant in water supplies because it creates potential public health hazards, unpleasant appearance, and operational difficulties. The most important of these is the potential public health hazard. The effectiveness of chlorine when used to disinfect water depends on the chlorine making contact with the bacteria. The suspended particles in turbid water can shelter bacteria from the chlorine. The bacteria may then travel through the treatment process, reach the consumer, and cause disease. Turbid water may also contain particles of organic matter, which can react with chlorine to form trihalomethanes—potentially harmful chlorinated organic compounds.

Every glass of drinking water is judged by its clarity. Obviously, cloudy or turbid water is unappealing. Turbidity in excess of 5 NEPHELOMETRIC TURBIDITY UNITS (NTU) is noticeable to the consumer, making drinking water very unappealing. This is dangerous because the consumer may either substitute a more appealing, but possibly unsafe, private water supply, or he may install some home treatment device that, in time, may harbor and release disease-causing organisms due to poor maintenance.

Turbidity analyses are also used to evaluate in-plant operations. Turbidity measurements after settling and before filtration will monitor performance of the coagulation/flocculation and sedimentation processes. A rise in turbidity after settling indicates that the coagulant application should be changed or that operational corrections must be made. Settled water before filtration should have a turbidity of less than 10 NTU. If water with high turbidity reaches the filter, it will cause high filter head losses and short filter runs. Changes in raw-water turbidity usually require that the coagulant dosage be changed. Any noticeable change in turbidity within the unit process should be an immediate warning that operational adjustments are necessary.

Turbidity analyses are also used to monitor finished-water quality for compliance with state and federal drinking water standards. Turbidity averaging greater than 1 NTU in finished water indicates an operational problem requiring corrective action at the treatment plant.

Sampling

Turbidity analysis should be conducted on samples collected from raw water, sedimentation basin effluent, filter effluent, and finished water. Figure 5-6 shows some typical turbidity sampling points. The USEPA drinking water regulations require that all community and non-community water systems using surface-water sources test daily for turbidity levels at the point (or points) of entry to the distribution system.

At least 100 mL of sample should be collected in a clean glass or polyethylene container. Samples should be shaken and analyzed immediately after collection because the turbidity can change if the sample is stored. It may sometimes be impractical to run the turbidity test immediately. In such cases, the sample should be stored in the dark for no longer than 24 hours.

All filter plants should keep a continuous record of finished-water turbidity. Continuous-reading turbidimeters with recorders should be installed on the filter effluent piping. The quality of filter effluent is then continuously determined, reported, and recorded. The turbidimeter signal can sound alarms to indicate the need to shut down an improperly operating filter. This increases

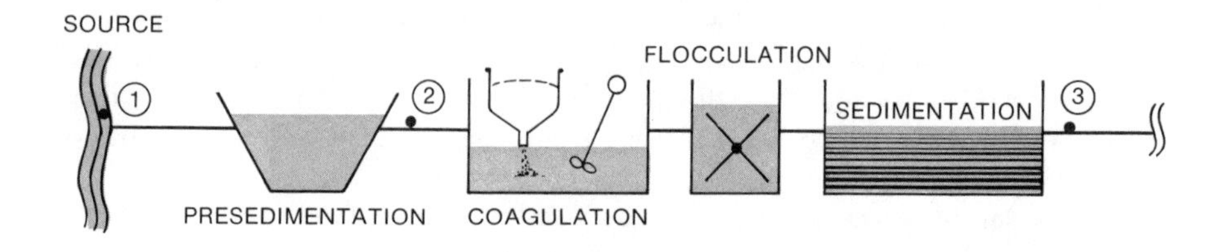

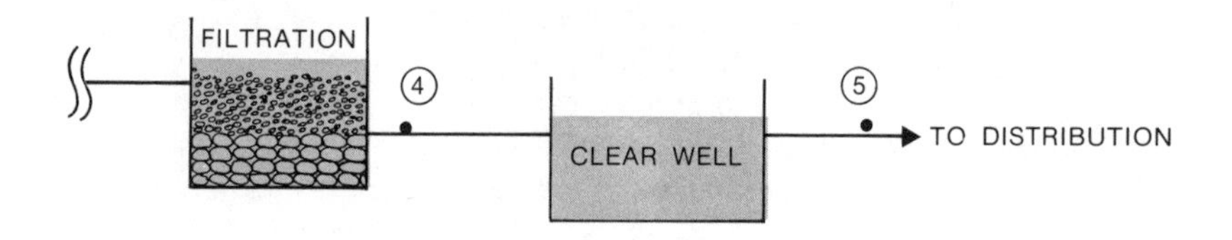

1 Turbidity of raw water entering the plant.

2 Turbidity reduction by presedimentation; help determine coagulant dose.

3 Turbidity removal by coagulation/flocculation and sedimentation processes; assist in determining operational changes.

4 Turbidity after filtration; continuous monitoring of turbidity for each filter to help determine filter efficiency and need for backwashing.

5 Turbidity of all treated water leaving the plant; monitor compliance with drinking water regulations.

Figure 5-6. Typical Turbidity Sampling Points

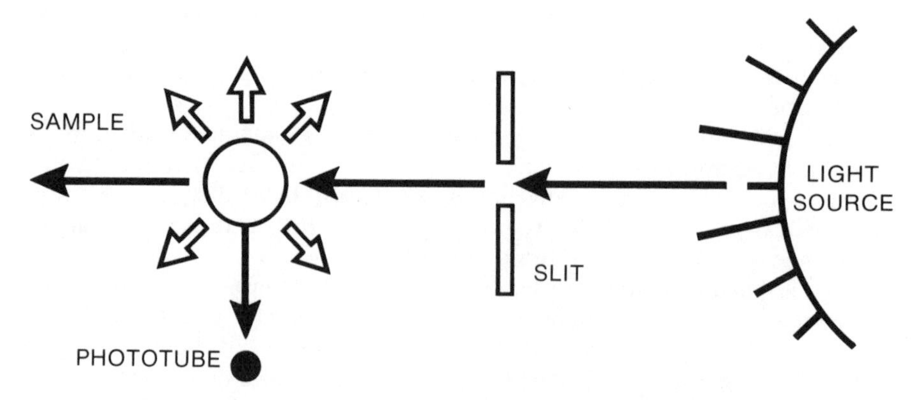

Figure 5-7. Scattering of Light by a Nephelometric Turbidimeter

Table 5-5. Scale of Nephelometric Turbidity Measurements

1000 NTU	Highly turbid surface water
100 NTU	Average for Missouri River
10 NTU	Average for lake water
1 to 5 NTU	Coagulated and settled water (filter influent)
0.1 NTU	Finished-water turbidity (filter effluent)

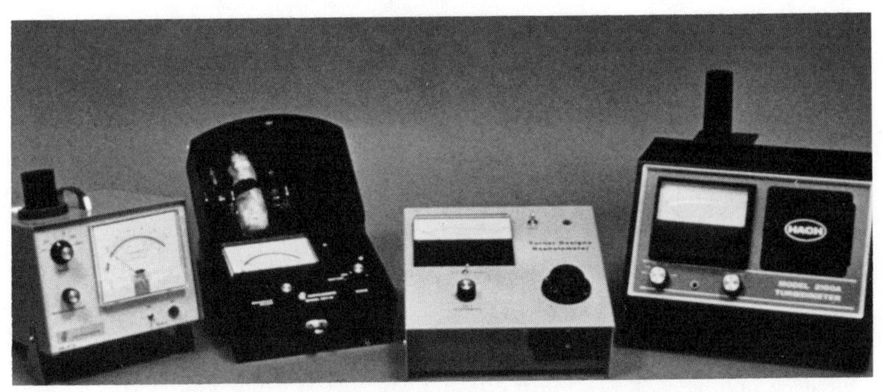

Figure 5-8. Commercially Available Nephelometers

the reliability of the filter operation and is especially important in assuring safe operation of pressure filters and high rate [4 to 6 gpm/sq ft (2.7 to 4.1 mm/s)] filter plants.

Methods of Determination

A conventional turbidimeter measures the passage, or transmission, of light. Measurements made with a conventional turbidimeter are expressed as JTUs (JACKSON TURBIDITY UNITS) or FTUs (FORMAZIN TURBIDITY UNITS).

The more modern NEPHELOMETRIC TURBIDIMETER measures the scattering of light (Figure 5-7). Measurements made with this unit are expressed as NTUs (nephelometric turbidity units). The USEPA drinking water regulations specify the use of a nephelometric turbidimeter for all required monitoring.

Analysis is quick and easy with the nephelometric method. Nephelometry is useful for in-plant monitoring, and results can be compared from plant to plant. This is an advantage to operators seeking performance information from other facilities.

Figure 5-8 shows several commercially available nephelometers; any one can be used for turbidity analysis. Typical turbidity values are shown in Table 5-5. General procedures for using a turbidimeter are given in Appendix E.

5-18. Inorganic Chemicals in General

All surface- and ground-water sources contain a variety of inorganic chemicals. A major source of the inorganic chemicals is geologic formations that the water contacts. Other sources include industrial discharges and agricultural runoff.

Significance

Inorganic chemicals can cause public health or aesthetic problems. The inorganic chemicals that are a health concern are covered by the USEPA primary drinking water regulations. Those causing aesthetic problems are listed in the secondary regulations (see Module 1, Drinking Water Standards).

Sampling

Sampling procedures used for inorganic chemicals are discussed in Module 2, Sample Collection, Preservation, and Storage.

Methods of Determination

With few exceptions (some of which are discussed in other sections of this module) analyses for inorganic chemicals are conducted using an ATOMIC ABSORPTION SPECTROPHOTOMETER (AA unit). More specifically, an AA unit is used to determine the concentration of metals. It is typically used by large water utilities, state agencies, and commerical laboratories since it is expensive and requires experienced chemists for its operation.

The sample is ASPIRATED into a flame where it is vaporized. A special light source emits light at a wavelength that is characteristic of the element being measured. The amount of light absorbed by the vapor is measured. This absorbance is directly proportional to the concentration of the element in the sample.

Nitrate concentrations are determined using a colorimetric technique and an ordinary spectrophotometer.

5-19. Organic Chemicals in General

All surface waters contain organic chemicals. These natural compounds, found in soil and vegetation, enter into the surface water. Agricultural runoff and industrial contamination can also contribute significant amounts of organic chemicals. Ground water can also contain organic compounds. Organic compounds in ground water are usually from man-caused contamination.

Significance

Over 700 different organic chemicals have been identified in drinking water. At one time, it was thought that these compounds only caused tastes and odors in drinking water. However, it has been found that many of these compounds are known to or suspected to have adverse effects on man, including cancer. As a result, many water systems now monitor for organic chemicals in addition to those listed in the USEPA drinking water regulations (see Module 1, Drinking Water Standards).

Sampling

Because of the wide variety of chemicals involved, it is difficult to establish a set sampling procedure. When sampling it is important to follow the instructions from the laboratory that will be doing the analysis.

Methods of Determination

The gas chromatograph (described under 5-16, Trihalomethanes) must be used for analysis of organic compounds. In most cases, the analysis requires use of very sophisticated equipment (a gas chromatograph-mass spectrometer, or GC-MS) and detailed chemical techniques. Only a few large water-utility laboratories are equipped to conduct these tests.

Selected Supplementary Readings

Standard Methods for the Examination of Water and Wastewater. APHA, AWWA, and WPCF. Washington, D.C. (15th ed., 1980).

Methods for Chemical Analyses of Water and Wastes. EPA-625/6-74-003. Ofce. of Technology Transfer (1974).

Simplified Procedures for Water Examination. AWWA Manual M12. AWWA, Denver, Colo. (1975).

Glossary Terms Introduced in Module 5

(Terms are defined in the Glossary at the back of the book.)

Anaerobic
Aspirate
Atomic absorption spectrophotometer
Atomic absorption spectrophotometric method
BOD
Breakpoint
Breakpoint chlorination
Buffering capacity
Calibrated
Chlorine demand
Color unit
Colorimetric method
Combined chlorine residual
CU
EDTA
Electrode method
Equilibrium
Formazin Turbidity Units

Free available chlorine residual
Gas Chromatograph
Gravimetric procedure
Indicator
Iodometric method
Iron bacteria
Jackson Turbidity Units
Langelier Saturation Index
Methyl orange
Modified Winkler Method
Mottled
NTU
Nephelometric turbidimeter
Nephelometric turbidity unit
Nomographic method
Oxidant
Oxidize
pH_s
Phenanthroline method
Phenolphthalein indicator
Platinum-cobalt method
Potentiometrically
Precipitate
Probe method
Recarbonation
SPADNS method
Selective adsorption
Sequestering
Stable
Strip
Tests
Threshold odor number
Titrimetric method
TON

Review Questions

(Answers to Review Questions are given at the back of the book.)

1. What is the importance of alkalinity in water treatment?

2. What method is used to determine alkalinity?

3. What information is needed to calculate the Langelier Index?

4. A water has the following chemical constituents:
   Alkalinity = 80 mg/L as $CaCO_3$
   Temperature = 12° C
   Total Dissolved Solids = 400 mg/L
   pH = 7.8
   Ca^{+2} = 200 mg/L as $CaCO_3$
   Determine the Langelier Saturation Index. Is this water stable? If not, what would be expected to happen?

5. Why should the Langelier Index be used with caution?

6. What are the two types of chlorine residual?

7. What is chlorine demand?

8. Eight mg/L chlorine is added at a water treatment plant and a free available concentration of 0.5 mg/L is determined as the water leaves the plant. What is the chlorine demand?

9. If the total available chlorine residual is 3 mg/L and the combined available chlorine residual is 2.6 mg/L, what is the free available chlorine residual concentration?

10. What form of chlorine residual is the most powerful disinfectant?

11. Name two methods for determining chlorine residual.

12. What is the major purpose for jar testing?

13. Jar tests were conducted on two different coagulants with the following results:

	Aluminum Sulfate	*Ferric Chloride*
Optimal dose	30 mg/L	20 mg/L
Turbidity at optimal dose	2 NTU	2 NTU
Pounds used/mil gal	250	176

   a. If aluminum sulfate costs $0.10/lb, what is the cost to treat one million gallons?
   b. If ferric chloride costs $0.20/lb, what is the cost to treat one million gallons?
   c. Which coagulant would cost the least to use and what is the savings per million gallons?

14. Why does surface water generally contain more color than ground water?

15. What useful purpose does dissolved oxygen serve in water treatment?

16. Which technique is preferred for DO analysis and why?

17. Why is fluoride added to water?

18. If fluoride is being added to water, what is the minimum testing frequency for fluoride in the treated water?

19. What problem might be created by a high carbon dioxide concentration in treated water?

20. What information is necessary for determining free carbon dioxide by the nomographic method?

21. Define hardness.

22. What method is used to determine hardness concentration in water? What chemical process is involved?

23. What problems are associated with iron and manganese in water supplies?

24. Assume a water plant's finished water has iron = 0.01 mg/L and manganese = 0.02 mg/L. At a consumer's tap, iron = 0.04 mg/L and manganese = 0.01 mg/L. What is a possible cause of this change in concentration?

25. Define pH.

26. Give examples of alkaline, acidic, and neutral pH readings.

27. What is the preferred method for pH determination?

28. Why are odor tests preferred over taste tests for drinking water?

29. What are the purposes of odor tests?

30. What type of sample bottle should be used for odor tests?

31. How does temperature affect chemical reactions?

32. What procedure is used to determine total filterable residue?

33. What are trihalomethanes?

34. What is added to the sample bottle to stop the THM reaction?

35. What is the recommended maximum storage time for samples that will be analyzed for turbidity?

36. What is the only method of turbidity analysis approved for monitoring under the USEPA drinking water regulations?

37. Are suspended solids the same as turbidity? Explain.

38. How are turbidity and bacteriological quality related?

39. The following data was taken from a water treatment plant practicing coagulation and flocculation:

    Raw water turbidity = 40 NTU
    Sedimentation effluent turbidity = 20 NTU
    Filter effluent turbidity = 5 NTU

    Are all processes working efficiently? Explain.

40. What is the principal technique for analysis of dissolved metals in water?

Study Problems and Exercises

1. You have been asked to develop a monitoring program to assist in achieving more effective operation of the water treatment plant. Treatment processes consist of presedimentation, addition of aluminum sulfate, rapid mix, flocculation, sedimentation, filtration, chlorination, and fluoridation.
   a. List the raw- and treated-water data necessary for effective operational control.
   b. What data within the plant are necessary? Where would you recommend the samples be taken?
   c. What laboratory equipment is required to conduct the tests in (a) and (b)?

Appendix A

Coliform Samples Required per Population Served

Table A.1. Coliform Samples Required Per Population Served*

Population Served	*Minimum Number of Samples per Month*	*Population Served*	*Minimum Number of Samples per Month*
25 to 1000 †	1	90,001 to 96,000	95
1000 to 2500	2	96,001 to 111,000	100
2501 to 3300	3	111,001 to 130,000	110
3301 to 4100	4	130,001 to 160,000	120
4101 to 4900	5	160,001 to 190,000	130
4901 to 5800	6	190,001 to 220,000	140
5801 to 6700	7	220,001 to 250,000	150
6701 to 7600	8	250,001 to 290,000	160
7601 to 8500	9	290,001 to 320,000	170
8501 to 9400	10	320,001 to 360,000	180
9401 to 10,300	11	360,001 to 410,000	190
10,301 to 11,100	12	410,001 to 450,000	200
11,101 to 12,000	13	450,001 to 500,000	210
12,001 to 12,900	14	500,001 to 550,000	220
12,901 to 13,700	15	550,001 to 600,000	230
13,701 to 14,600	16	600,001 to 660,000	240
14,601 to 15,500	17	660,001 to 720,000	250
15,501 to 16,300	18	720,001 to 780,000	260
16,301 to 17,200	19	780,001 to 840,000	270
17,201 to 18,100	20	840,001 to 910,000	280
18,101 to 18,900	21	910,001 to 970,000	290
18,901 to 19,800	22	970,001 to 1,050,000	300
19,801 to 20,700	23	1,050,001 to 1,140,000	310
20,701 to 21,500	24	1,140,001 to 1,230, 000	320
21,501 to 22,300	25	1,230,001 to 1,320,000	330
22,301 to 23,200	26	1,320,001 to 1,420,000	340
23,201 to 24,000	27	1,420,001 to 1,520,000	350
24,001 to 24,900	28	1,520,001 to 1,630,000	360
24,901 to 25,000	29	1,630,001 to 1,730,000	370
25,001 to 28,000	30	1,730,001 to 1,850,000	380
28,001 to 33,000	35	1,850,001 to 1,970,000	390
33,001 to 37,000	40	1,970,001 to 2,060,000	400
37,001 to 41,000	45	2,060,001 to 2,270,000	410
41,001 to 46,000	50	2,270,001 to 2,510,000	420
46,001 to 50,000	55	2,510,001 to 2,750,000	430
50,001 to 54,000	60	2,750,001 to 3,020,000	440
54,001 to 59,000	65	3,020.001 to 3,320,000	450
59,001 to 64,000	70	3,320,001 to 3,620,000	460
64,001 to 70,000	75	3,620,001 to 3,960,000	470
70,001 to 76,000	80	3,960,001 to 4,310,000	480
76,001 to 83,000	85	4,310,001 to 4,690,000	490
83,001 to 90,000	90	More than 4,690,001	500

*Source: EPA

†A community water system serving 25 to 1000 persons, with written permission from the state, may reduce this sampling frequency, except that in no case shall it be reduced to less than one per quarter. The decision by the state will be based on a history of no coliform bacterial contamination for that system and on a sanitary survey by the state showing the water system to be supplied solely by a protected ground-water source, free of sanitary defects.

Appendix B

Recommendation for Sampling and Preservation of Samples

Table B.1 Recommendation for Sampling and Preservation of Samples According to Measurement

Constituent	*Volume Required* mL	*Container**	*Preservative†*	*Holding Time‡*
Bacteriological				
Coliform	100	P,G	Cool, 4°C	Analyze as soon as possible
Physical Properties				
Color	500	P,G	Cool, 4°C	24 hours
Hardness	100	P,G	Cool, 4°C; HNO_3 to pH <2§	6 months
Odor	500	G only	Cool, 4°C	24 hours
pH	25	P,G	Determine on site	2 hours
Taste	500	G only	Cool, 4°C	24 hours
Total Dissolved Solids	100	P,G	Cool, 4°C	7 days
Temperature	1000	P,G	Determine on site	No holding
Turbidity	100	P,G	Cool, 4°C; store in dark	24 hours
Metals				
Total (dissolved and undissolved)	100 (for each metal)	P,G	HNO_3 to pH <2	6 months**
Total Mercury	500	P,G	HNO_3 to pH <2	28 days (glass) 13 days (hard plastic)
Inorganics, Non-Metallics				
Acidity	100	P,G	Cool, 4°C	24 hours
Alkalinity	200	P,G	Cool, 4°C	24 hours
Chlorine	500	P,G	Determine on site	No holding
Fluoride	300	P,G	None required	28 days

Table B.1 Recommendation for Sampling and Preservation of Samples According to Measurement (continued)

Constituent	*Volume Required* mL	*Container**	*Preservative†*	*Holding Time‡*
Inorganics, Non-Metallics (continued)				
Nitrate	100	P,G	Cool, 4°C; add H_2SO_4 to pH <2	48 hours
Carbon Dioxide	100	P,G	Analyze immediately	No holding
Dissolved Oxygen				
Electrode	300	G only	Determine on site	No holding
Winkler	300	G only	Fix on site	4–8 hours
Sulfate	50	P,G	Cool, 4°C	28 days
Organics**				
Pesticides and Herbicides	2000	G	Cool, 4°C	7 days
Trihalomethanes	25–250	G	Cool, 4°C	Analyze as soon as possible

*Plastic (P) or Glass (G). For metals, polyethylene with a polypropylene cap (no liner) is preferred.

†If the sample is stabilized by cooling, it should be warmed to 25° C for reading, or temperature correction made and results reported at 25° C.

‡It should be pointed out that holding times listed above are recommended for properly preserved samples based on currently available data. It is recognized that for some sample types, extension of these times may be possible while for other types, these times may be too long. Where shipping regulations prevent the use of the proper preservation technique or the holding time is exceeded, such as the case of a 24-hour composite, the final reported data for these samples should indicate the specific variance.

§Where HNO_3 cannot be used because of shipping restrictions, the sample may be initially preserved by icing and immediately shipped to the laboratory. Upon receipt in the laboratory, the sample must be acidified to a pH <2 with HNO_3 (normally 3 mL 1:1 HNO_3/litre is sufficient). At the time of analysis, the sample container should be thoroughly rinsed with 1:1 HNO_3 and the washings added to the sample (volume correction may be required).

**Sample bottles should be sealed with bottle caps that are teflon-lined.

Appendix C

Procedures for DPD Testing

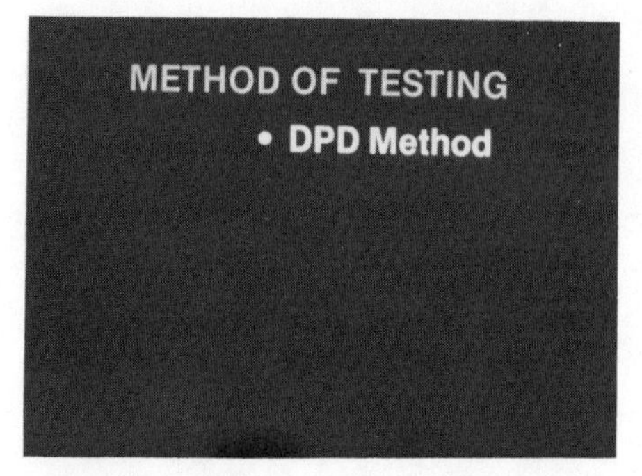

In measuring chlorine residual, only one method is approved for use under the USEPA drinking water regulations; i.e., the DPD Method. This method is a simple and quick way to measure chlorine residual. It takes less than 5 min to complete the test.

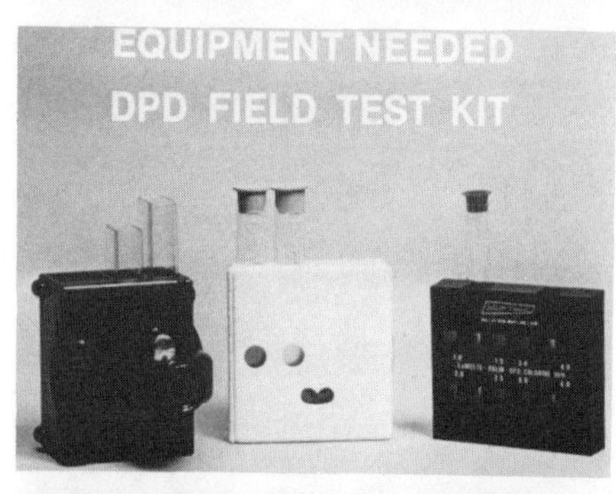

The only equipment needed is a DPD Field Test Kit. The necessary chemicals are provided with the kit.

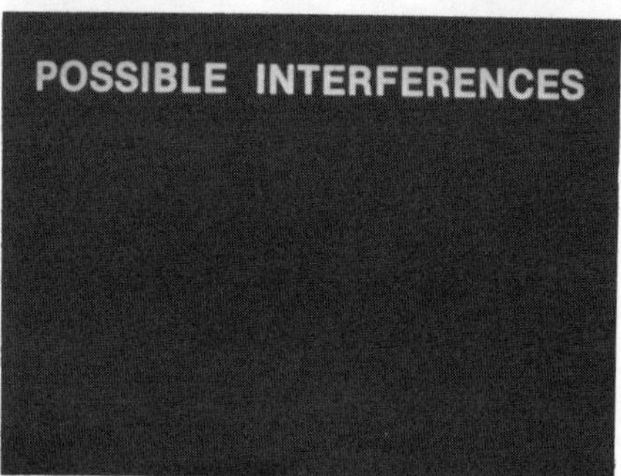

Before starting the chlorine residual test, consider those factors that can interfere with good test results.

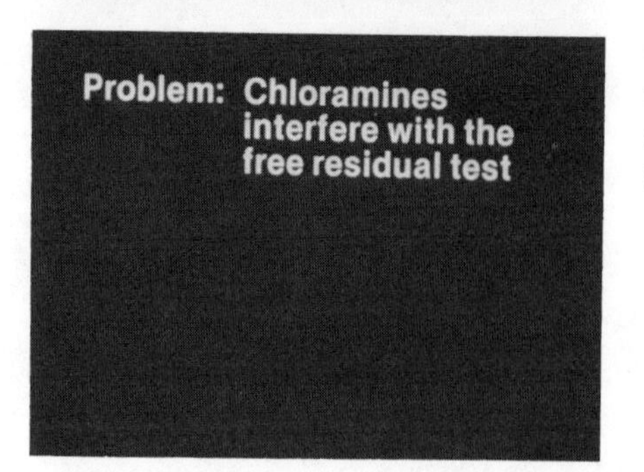

Chloramines, or the combined chlorine residual, can interfere or change the color of the free residual. To prevent this. . .

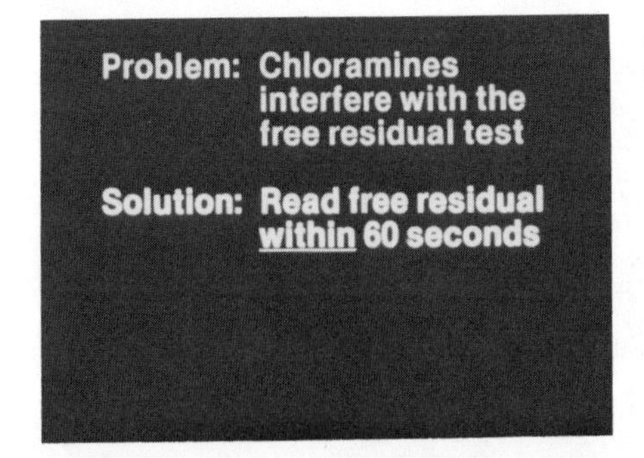

. . . read the free residual within 60 sec after adding the chemical.

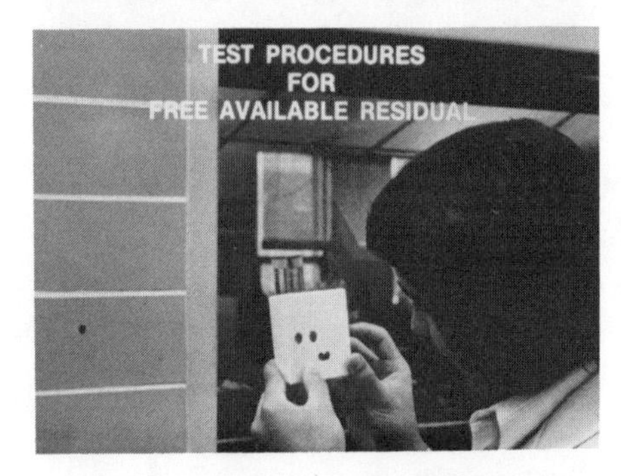

The test procedure for free available chlorine residual is a five step process.

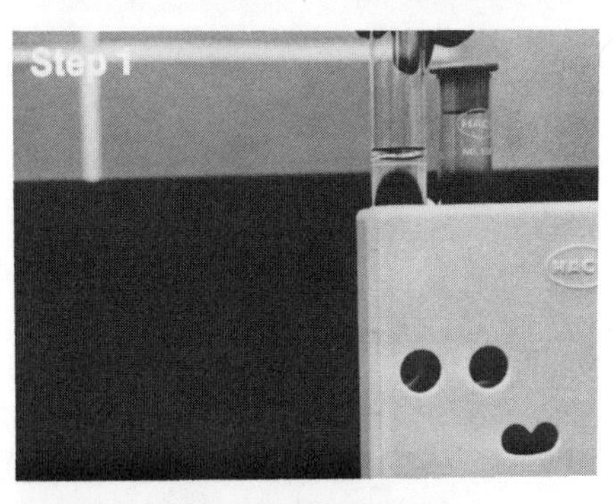

Fill one sample tube with sample water and place it in the holder behind the color comparator.

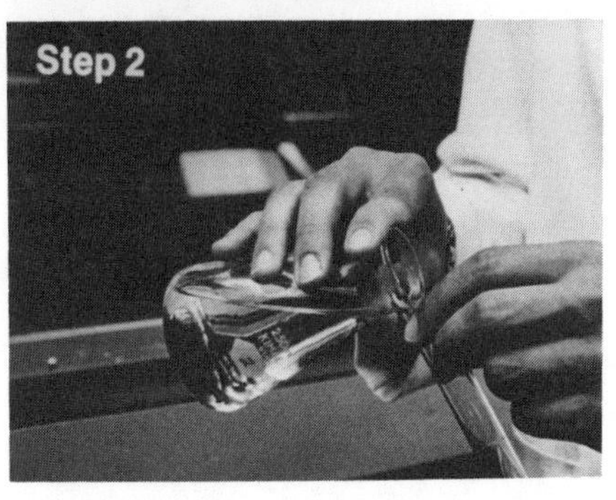

Fill the second tube to the 5 mL mark with the water to be tested.

Now add the free chlorine chemical to the second tube and gently swirl to mix.

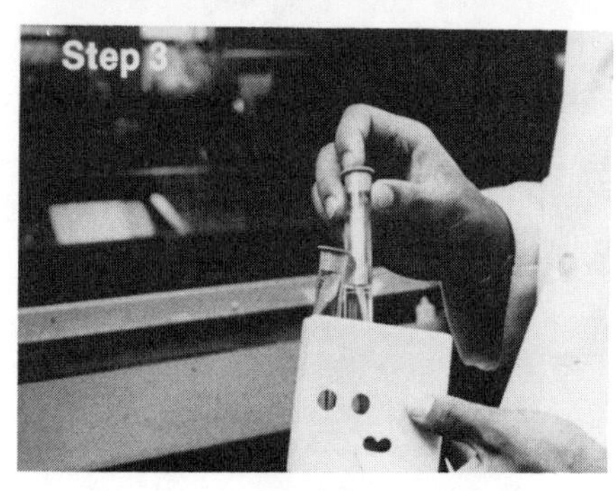

Then place the tube in the comparator.

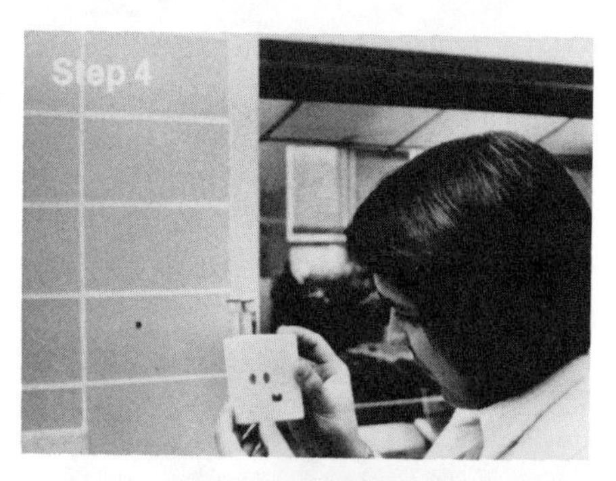

Hold the comparator up to natural light, rotating the color comparator wheel until the wheel color matches the sample color.

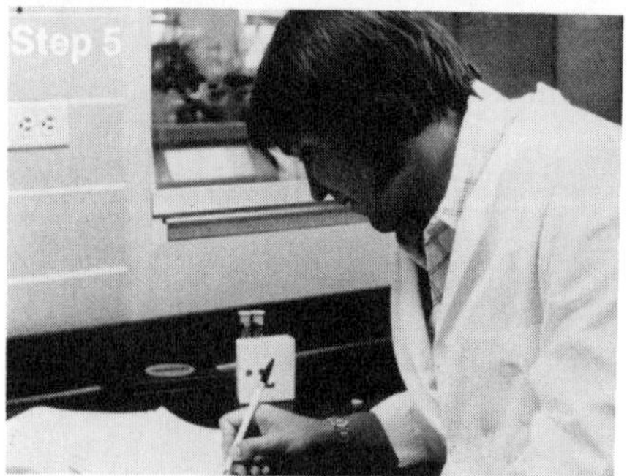

Read the residual in mg/L immediately, and record your results as mg/L of free available chlorine residual.

Appendix D
Jar Test Procedure

Jar Test Procedure

The following covers the basic jar test procedure. For additional information on variations of this procedure and practical applications of the test, refer to AWWA Manual M12, *Simplified Procedures for Water Examination.*

The purpose of the jar test is to determine such experimental conditions as the speed of the stirring and the length of the flash mix, flocculation, and settling intervals. The various conditions described in the following procedure are a good starting point for a laboratory trying the test for the first time. Some or all of the times and speeds may have to be changed to reflect the deficiencies of an old plant or the improved operation of a new or remodeled plant.

Since temperature plays an important role in coagulation, the raw-water samples should be collected and measured only after all other preparations have been made. This reduces the effect that room temperature might have on the sample.

Even the smallest detail may have an important influence on the result of a jar test. Therefore, all samples in a series of tests should be handled as nearly alike as possible.

Apparatus

1. *Stirring machine.* The machine has three to six paddles, capable of operating at variable speeds (from 0 to 100 rpm). Multiple stirring units are commercially available and generally superior to similar homemade outfits, which may require considerable construction time by a mechanic.
2. *Floc illuminator.* This item, located at the base of the laboratory stirrer, enables observation of small floc particles.
3. *Beakers.* Beakers should have a 1500-mL capacity. They should be short, large diameter beakers made of pyrex. Beakers of 500-mL or other capacity can be used with stirring machines of smaller size.
4. *Plastic pail.* The pail should be the household type, with a capacity in excess of 2 gal, for collecting the sample.
5. *Graduated cylinder.* The cylinder should have a 1000-mL capacity. A 500-mL graduate can be used for measuring samples into smaller-size beakers.
6. *Measuring pipets.* Pipets may have 1-, 5-, and 10-mL capacity, all graduated in 0.1-mL steps, for dosing samples rapidly with coagulant, suspensions, and other necessary solutions. These pipets should be rinsed thoroughly with tap water or distilled water to prevent caking with coagulant, suspensions, or other solutions being used.

7. *100-mL pipet.* This is used for withdrawing the coagulated and softened sample.
8. *Other.* Apparatus for determining color, turbidity, pH, and phenolphthalein and total alkalinity are also needed.

Coagulant Dosing Solution

The common coagulants are: aluminum sulfate (also called filter alum), $Al_2(SO_4)_3 \cdot 14H_2O$; ferrous sulfate (also called copperas), $FeSO_4 \cdot 7H_2O$; ferric sulfate (also called ferrisul), $Fe_2(SO_4)_3$; and sodium aluminate, $NaAlO_2$.

Dosing solutions or suspensions should be prepared from the stock materials actually used in plant treatment. Distilled water used for the preparation of lime suspensions should be boiled for 15 min to expel the carbon dioxide and then cooled to room temperature before the lime is added.

Steps to follow:

1. Weigh 10.0 g of material. Dissolve or suspend in 1 L of distilled water. Record the date of preparation on the bottle label.
2. Shake the suspension immediately before use.
3. Each 0.1 mL of this solution or suspension represents a dosage of 1 mg/L when added to a 1-L water sample, while each 1.0 mL of dosing solution represents a dosage of 10 mg/L in 1 L of sample. (See the following section for the test procedure to follow when using this suspension.)

Note: If 17.1 g of material is used in step 1, each 1.0 mL of the resulting solution represents a dosage of 1 grain per US gallon when added to 1 L of sample. If 14.3 g is used in step 1, each 1.0 mL represents 1 grain per Imperial gallon when added to 1 L of sample.

Test Procedure for Coagulation Treatment

Steps to follow:

1. Rinse six 1500-mL beakers with tap water and let the beakers drain for a few minutes in an upside-down position. Beakers that have had several days of use should be scrubbed inside and out with a brush and a household dishwashing detergent, finishing with a thorough rinse of tap water.
2. Clean the stirring machine paddles with a damp cloth.
3. Collect a sample of raw water and complete steps 4 through 8 within 20 min. Throw away the sample and collect a fresh sample if the work must be interrupted during this critical stage. Otherwise, the settling of high turbidity and an increase in the sample temperature from the heat of the laboratory may cause erroneous results.
4. Stand the beakers right side up and pour more than 1 L of raw water into each one.
5. Taking one beaker at a time, pour some of the raw water back and forth between the beaker and a 1-L graduated cylinder. Finally, fill the graduated cylinder to the 1-L mark and discharge the excess raw water in the beaker. Return the measured 1-L sample to the beaker.

6. Place all the beakers containing the measured 1-L samples on the stirring machine.
7. With a measuring pipet, add increasing doses of coagulant solution as rapidly as possible. Select a series of doses so that the first beaker will represent undertreatment and the last beaker will represent overtreatment. When a proper series is set up, the succession of beakers will show poor, fair, good, and excellent coagulation at the end of the run. It will be necessary to repeat the jar test once or twice to determine the proper series of doses for the desired results.
8. Lower the stirring paddles into the beakers, start the stirring machine, and operate it for 1 min at a speed of 60 to 80 rpm.
9. Reduce the stirring speed over the next 30 sec to 30 rpm and continue stirring at that speed for exactly 15 min.
10. Observe each beaker for the appearance of "pinpoint" floc and record the time and order of such appearance.
11. Stop the stirring machine and allow the samples to settle for 5, 15, 30, or 60 min. Observe the floc and record the order of settling. Describe the results as poor, fair, good, or excellent. A hazy sample indicates poor coagulation. Properly coagulated water contains floc particles that are well-formed, and the liquid between the particles is clear. The lowest coagulant dosage that brings down the turbidity during the jar test should first be tried in plant operation.
12. Using a 100-mL pipet, withdraw a portion of the top 1.5 in. of sample from each beaker.
13. Determine the color, turbidity, pH, and phenolphthalein and total alkalinity of the coagulated sample according to the directions given in *Standard Methods for the Examination of Water and Wastewater* or AWWA Manual M12, *Simplified Procedures for Water Examination.*

Appendix E

Procedures for Using a Turbidimeter

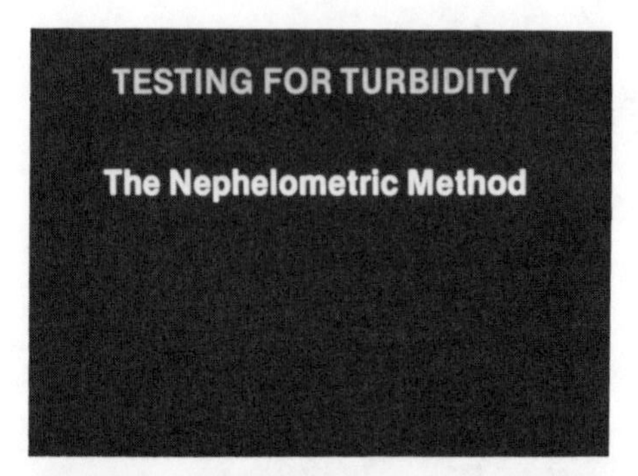

Although there are two standard methods for measuring turbidity, only one, the nephelometric method, is approved for use under the USEPA drinking water regulations.

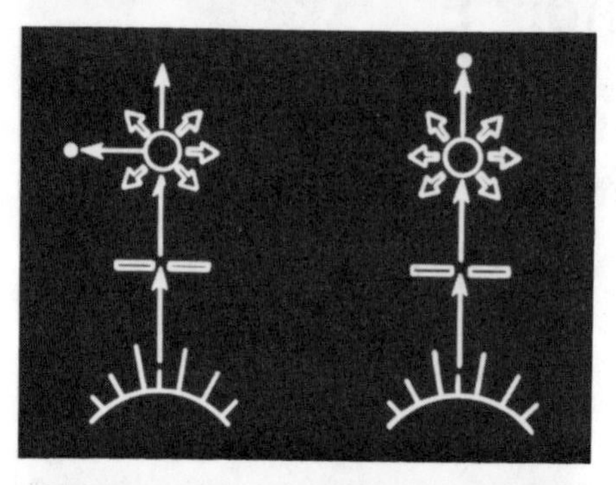

The nephelometric method measures the amount of light scattered by the turbidity particles. It was selected because it can consistently and repeatedly measure turbidities in the drinking water range—0 to 5 NTUs.

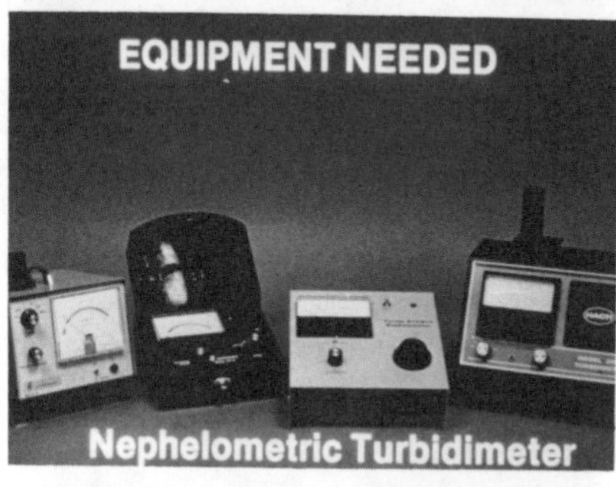

The only piece of equipment you will need is a nephelometric turbidimeter, together with standard turbidity solutions.

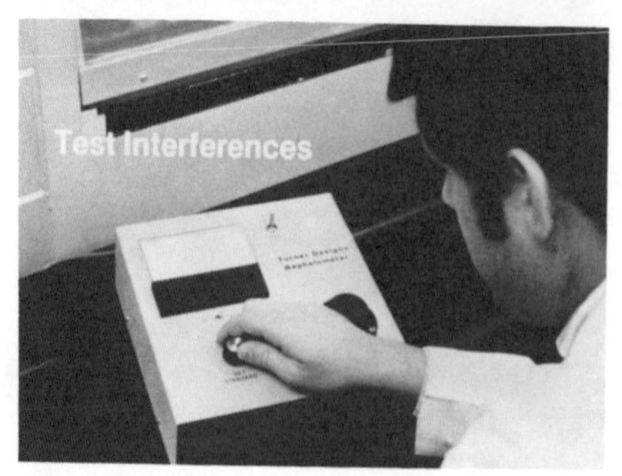

Before running the turbidity test, review those factors that can interfere with good test results.

Debris And Sediment

Floating debris and rapid settling particles will cause artificially low readings. If this occurs . . .

Retake The Sample

. . . retake the sample.

Dirty Or Scratched Tubes

Dirty, scratched, or chipped sample tubes can cause high readings. To prevent this . . .

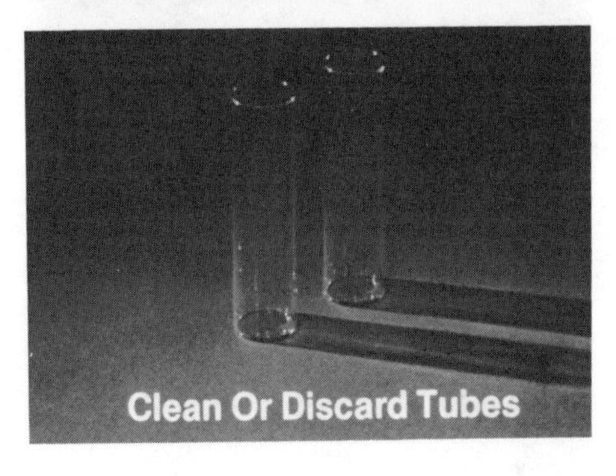
Clean Or Discard Tubes

. . . clean the tubes and discard those that are scratched, chipped, or cracked.

Air bubbles in the sample will cause high readings. If this occurs . . .

. . . wait for the air bubbles to clear.

Vibration can cause high readings.

For this reason, it is important to locate the turbidimeter on a sturdy bench. The bench should be on solid footing, such as a concrete ground-level floor. Do not touch the bench during the test.

TEST INTERFERENCES

- **Floating or settling debris**
- **Dirty or scratched sample tubes**
- **Air Bubbles**
- **Vibration**

These are the four factors that interfere with good test results.

TURBIDITY TEST PROCEDURE

- **Calibration**
- **Measurement**

The turbidity test procedure is a two-step process.

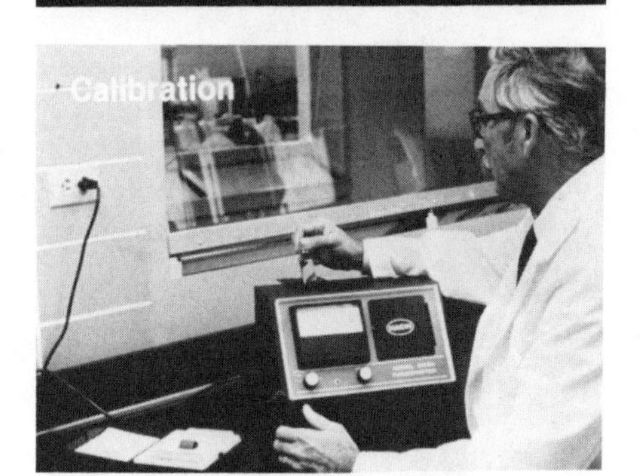

To calibrate the turbidimeter, insert a tube containing a standard suspension of known turbidity.

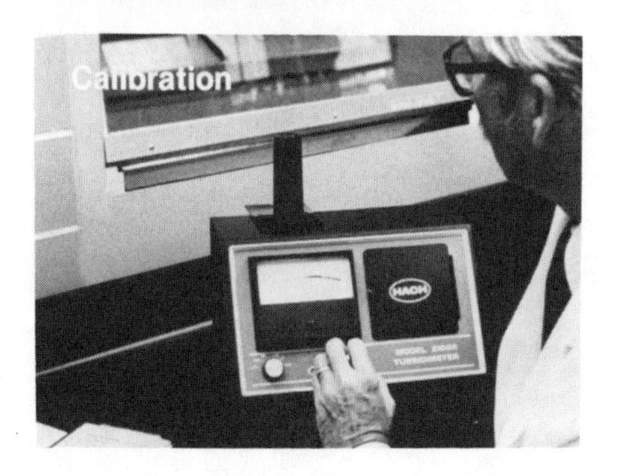

Adjust the turbidimeter needle until it registers the known value.

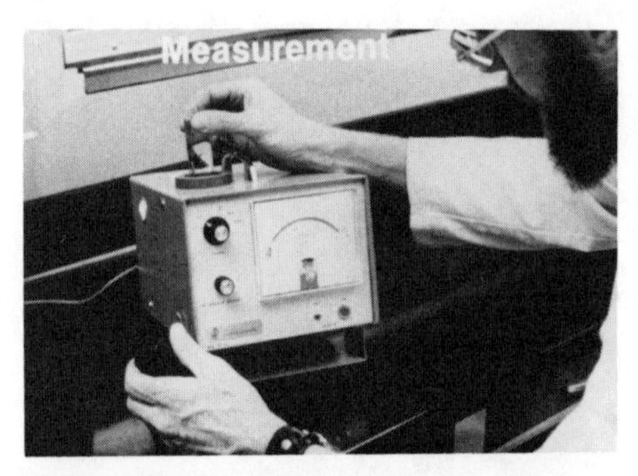

Now remove the standard, insert your sample . . .

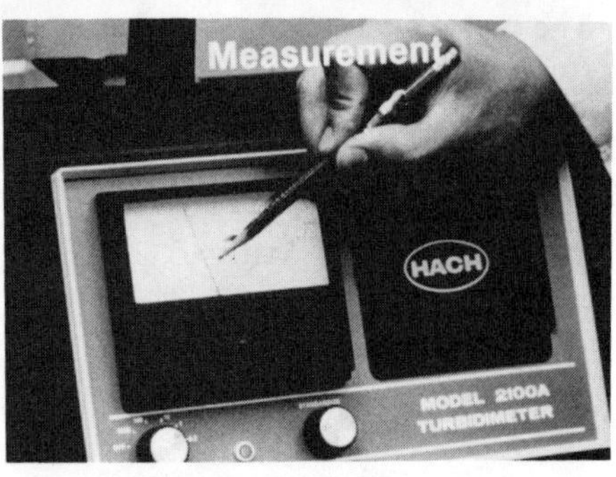

. . . and read and record the turbidity value directly from the instrument.

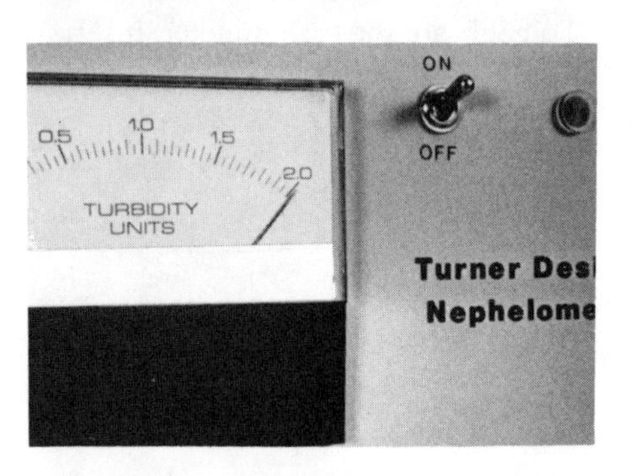

If the turbidity of the sample is above the range of your instrument, as it might be in a raw-water sample, you will first have to dilute the sample until the turbidity falls within the instrument range.

For example, take 50 mL of sample and add 50 mL of turbidity-free or distilled water.

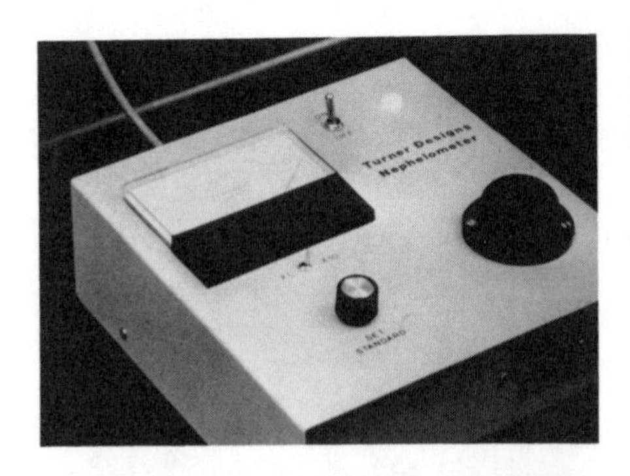

Now measure the turbidity. If it is still above the range, add another 50 mL of turbidity-free water.

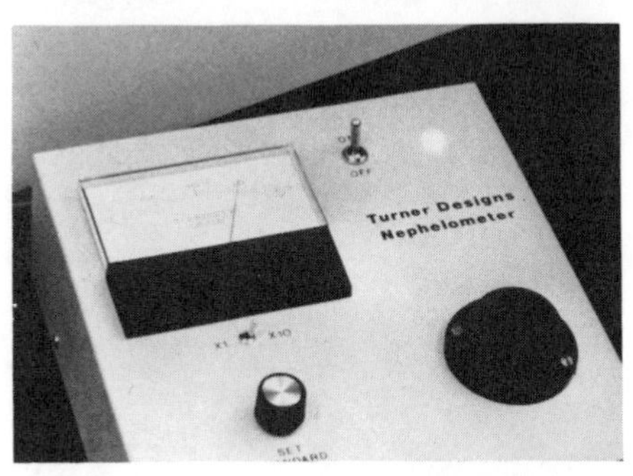

Keep measuring the turbidity until the measurement falls within the range of the instrument. Now . . .

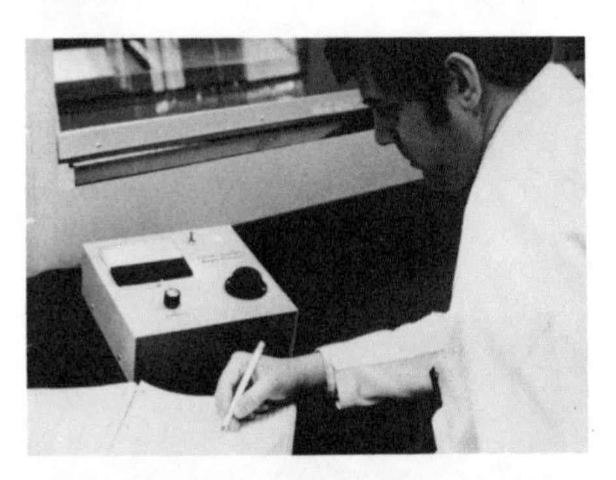

. . . read and record the turbidity measurement and record the amount of dilution water added.

If turbidity was measured directly from an undiluted sample, no calculations are needed. However, if the sample was diluted, use the following to calculate turbidity values.

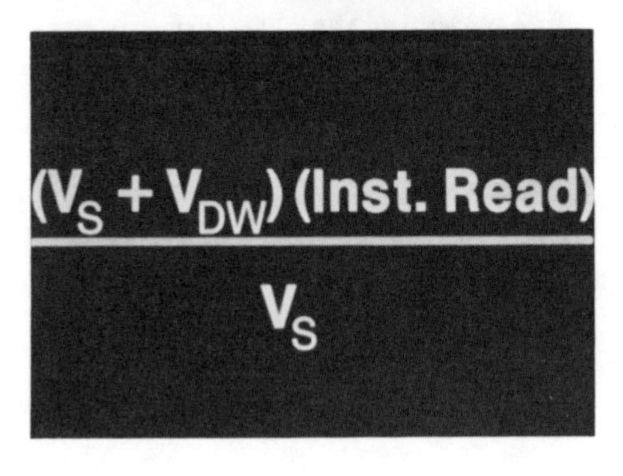

The equation tells you to add the volume of the sample to the volume of the dilution water, multiply that number by the instrument reading, then divide that result by the sample volume. Here is how it works.

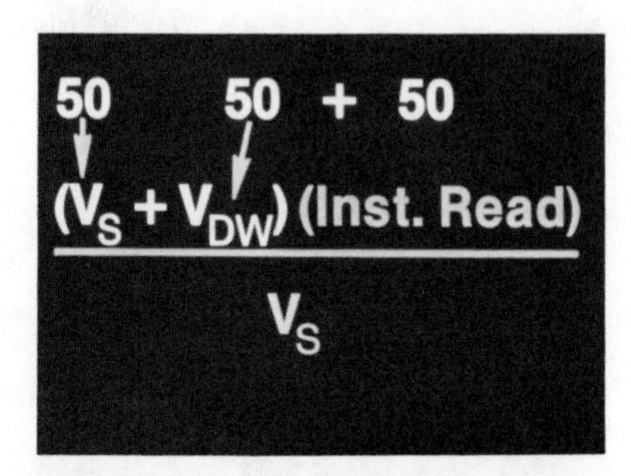

In our example, we started with 50 mL of sample. We then added dilution water twice, adding 50 mL each time.

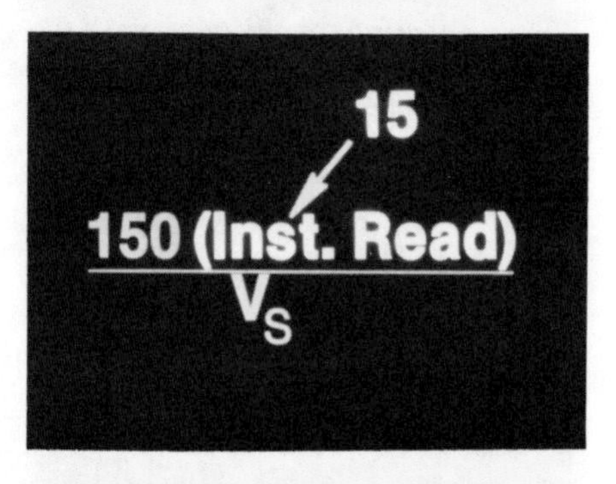

The diluted water sample produced an instrument reading of 15.

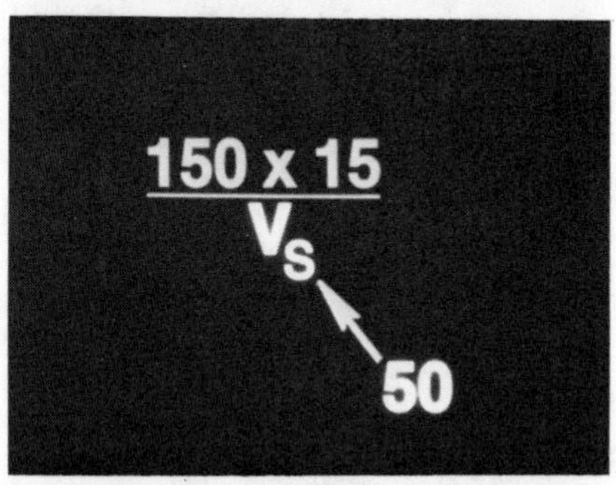

Next, fill in the undiluted sample volume—50 mL.

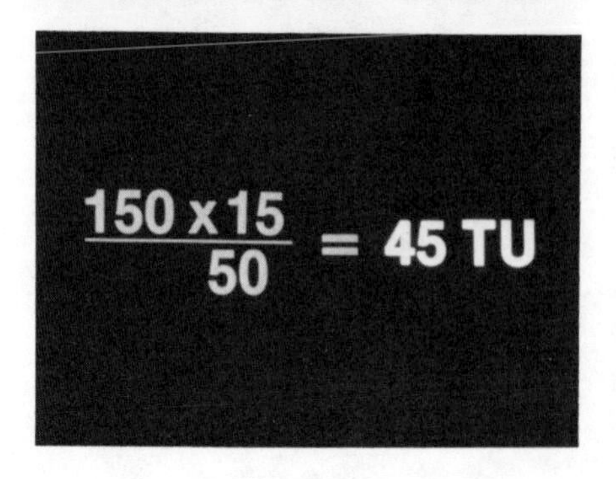

By performing the necessary arithmetic, you can calculate the turbidity of the original undiluted sample.

Glossary

Glossary

Words defined in the glossary are set in SMALL CAPITAL LETTERS where they are first used in the text.

Agar, 74 A nutrient preparation used to grow bacterial colonies in the laboratory. Agar is poured into PETRI DISHES to form agar plates or into CULTURE TUBES to form agar slants.

Anaerobic, 91, 95, 99 The absence of air or free oxygen.

Analytical balance, 60 A sensitive BALANCE used to make precise weight measurements.

Aspirate, 108 To remove a fluid from a container by suction.

Aspirator, 58 A T-shaped plumbing fixture connected to a water faucet. It creates a partial vacuum for filtering operations.

Atomic absorption spectrophotometer, 63,108 A SPECTROPHOTOMETER used to determine the concentration of metals in water and other types of samples.

Atomic absorption spectrophotometric method, 96, 97 An analytical technique used to identify the constituents of a sample by detecting which frequencies of light the sample absorbs.

Autoclave, 55 A device that sterilizes laboratory equipment by using pressurized steam.

Autoclaved, 44 Sterilized with steam at elevated temperature and pressure.

BOD (biochemical oxygen demand), 92 A measure of the amount of oxygen used in the biochemical oxidation of organic matter over a specified time (usually 5 days) and at a specific temperature (usually 35°C). Used to indicate the level of contamination in water or contamination potential of a waste.

Bacterial aftergrowth, 77 Growth of bacteria in treated water after it reaches the distribution system.

Balance, 60 An instrument used to measure weight.

Beaker, 44 A container with an open top, vertical sides, and a pouring lip used for mixing chemicals.

Borosilicate glass, 44 A type of heat-resistant glass used for labware.

Breakpoint, 86 The point at which the chlorine dosage has satisfied the CHLORINE DEMAND.

Breakpoint chlorination, 86 The addition of chlorine to water until the CHLORINE DEMAND has been satisfied and free chlorine residual is available for disinfection.

Buffering capacity, 82 The capability of water or chemical solution to resist a change in pH.

Buret, 44 Graduated glass tube fitted with a stopcock. Used to dispense solutions during TITRATION.

Burner, 59 A high-temperature heating device that uses natural or bottled gas. Also called a Bunsen burner.

CU, 90 See COLOR UNIT.

Calibrate, 98 To adjust a measuring instrument so that it gives the correct result with a known concentration or sample.

Check sampling, 9 Samples taken to verify ROUTINE SAMPLES or initial samples. The NIPDWRs require check sampling when routine sampling indicates a violation.

Chlorine demand, 87 The quantity of chlorine consumed by reaction with substances in the water.

Coliform (total coliform), 73 See COLIFORM BACTERIA.

Coliform bacteria (also called coliform-group bacteria), 5 A group of bacteria predominantly inhabiting the intestines of man or animal, but also occasionally found elsewhere. Presence of the bacteria in water is used as an indication of fecal contamination (contamination by human or animal wastes).

Colony counter, 50 An instrument used to count bacterial colonies for the STANDARD PLATE COUNT test.

Color comparator, 62 A device used for tests such as chlorine residual or pH. Concentrations of constituents are determined by visual comparison of a permanent standard (usually sealed in glass or plastic) and a water sample.

Color unit, 90 The unit of measure used to express the color of a water sample.

Colorimeter, 62 An instrument that measures the concentration of a constituent in a sample by measuring the intensity of color in that sample. The color is usually created by mixing a chemical reagent with the water sample according to a specific test procedure.

Colorimetric method, 98 Any analytical method that measures a constituent in water by determining the intensity of color. The color is usually produced when a chemical solution specified by the particular procedure is added to the water.

Combined chlorine residual, 86 The chlorine residual produced by the reaction of chlorine with substances in the water. Since the chlorine is "combined" it is not as effective a disinfectant as free chlorine residual.

Community system, 3 As defined by the NIPDWRs, a PUBLIC WATER SYSTEM that serves at least 15 service connections used by year-round residents or regularly serves at least 25 year-round residents.

Completed test, 73 The third major step of the MULTIPLE-TUBE FERMENTATION test. This confirms that positive results from the PRESUMPTIVE TEST are due to COLIFORM BACTERIA.

Composite sample, 25 A series of individual or GRAB SAMPLES taken at different times from the same sampling point and mixed together.

Compound microscope, 66 A MICROSCOPE with two or more lenses.

Confirmed test, 73 The second major step of the MULTIPLE-TUBE FERMENTATION test. This confirms that positive results from the PRESUMPTIVE TEST are due to COLIFORM BACTERIA.

Culture tube, 49 A hollow, slender, glass tube with an open top and a rounded bottom used in microbiological testing procedures such as the MULTIPLE-TUBE FERMENTATION test.

Deionizer, 57 A device used to remove all dissolved inorganic ions from water.

Deluge shower, 56 A safety device used to wash chemicals off the body quickly.

Dessicator, 51 A tightly sealed container used to cool heated items before they are weighed. This prevents the items from picking up moisture in the air and increasing in weight.

Dilution bottle, 45 Heat-resistant glass bottle used for diluting bacteriological samples before analysis. Also called MILK DILUTION BOTTLE or FRENCH SQUARE.

Double-pan balance, 60 A BALANCE that weighs material by counter balancing material placed on one pan with brass weights placed on the other pan.

EDTA (ethylenediaminetetraacetic acid), 95 A chemical used to SEQUESTER, or tie up, calcium and magnesium ions. Used in the hardness test.

Electrode method, 92, 93 Any analytical procedure that uses an electrode connected to a millivoltmeter to measure the concentration of a constituent in water.

Electrophotometer, 63 A PHOTOMETER that uses different colored glass filters to produce the desired wavelengths for analyses. Also called a filter photometer.

Equilibrium, 91 A balanced condition in which the rate of formation and the rate of consumption of a constituent or constituents are equal.

Erlenmeyer flask, 45 Bell-shaped container used for heating and mixing chemicals and culture media.

Evaporating dish, 48 Glass or porcelain dish in which samples are evaporated to dryness using high heat.

Eye wash, 56 A safety device used to wash chemicals from the eyes. The device, which resembles a drinking fountain, directs a gentle spray of water into each eye.

Filter (laboratory), 59 A porous layer of paper, glass fiber, or cellulose acetate used to remove particulate matter from water samples and other chemical solutions.

Filter paper, 59 Paper with pore size usually between 5 and 10 μm used to clarify chemical solutions, collect particulate matter, and separate solids from liquids.

Filtering crucible, 48 A small porcelain container with holes in the bottom, used in the total suspended solids test. Also known as a Gooch crucible.

Flaming, 37 Passing a flame over the end of a faucet in order to kill bacteria before taking a water sample for bacteriological sampling. The procedure is no longer recommended because it may damage the faucet.

Flask, 45 A container, often narrow at the top, used for holding liquids. There are many types of flasks, each with its own specific name and use.

Flow-proportional composite, 25 A COMPOSITE SAMPLE in which individual sample volumes are proportional to flow rate at the time of sampling.

Formazin Turbidity Unit, 107 Turbidity unit obtained when a chemical solution of formazin is used as a standard to CALIBRATE the TURBIDIMETER. If a NEPHELOMETRIC TURBIDIMETER is used, NEPHELOMETRIC TURBIDITY UNITS and formazin turbidity units are equivalent.

Free available chlorine residual, 86 The residual formed once all the CHLORINE DEMAND has been satisfied. The chlorine is not combined with other constituents in the water and is "free" to kill microorganisms.

French square, 45 See DILUTION BOTTLE.

Full-face shield, 55 A shatter-proof plastic shield worn to protect the face from flying particles and chemicals.

Fume hood, 51 A large enclosed cabinet equipped with a fan to vent fumes from the laboratory. Mixing and heating of chemicals are done under the hood to prevent fumes from spreading through the laboratory.

Funnel, 46 A utensil used in the laboratory for pouring liquids into FLASKS and other containers. Laboratory funnels are either glass or plastic.

Gas chromatograph, 103 An instrument used to measure the concentration of organic compounds in water.

Glass-fiber filter, 59 Filters made of uniform glass fibers with a pore size from 0.7 to 2.7 μm. Used to filter fine particles and algae while maintaining a high flow rate.

Gooch crucible, 48 See FILTERING CRUCIBLE.

Grab sample, 24 A single water sample collected at one time from a single point.

Graduated cylinder, 46 Tall, cylindrical, glass or plastic container with a hexagonal base and a pouring lip. Used for measuring liquids quickly without great accuracy. Graduations are marked on the side.

Gravimetric procedure, 102 Any analytical procedure that uses the weight of a constituent to determine its concentration.

Gross alpha activity, 9 Radioactivity due to materials that emit alpha particles such as radium. An alpha particle is positively charged and consists of two neutrons and two protons.

Herbicide, 5 A compound, usually a synthetic ORGANIC CHEMICAL, used to stop or retard plant growth.

Hot plate, 59 An electrical heating unit used to heat solutions.

Incubate, 74 To maintain microorganisms at a temperature and in an environment favorable to their growth.

Incubator, 52 A heated container that maintains a nearly constant temperature for development of MICROBIOLOGICAL cultures.

Indicator, 95 A chemical solution used to produce a visible change, usually in color, at a desired point in a chemical reaction, generally a prescribed end point.

Indicator organism, 72 Microorganism whose presence or absence indicates the presence or absence of fecal contamination in water.

Inorganic chemical, 5 A chemical substance of mineral origin not having carbon in its molecular structure.

Insecticide, 5 A compound, usually a synthetic ORGANIC CHEMICAL, used to kill insects.

Iodometric method, 92 A procedure for determining the concentration of dissolved oxygen in water. Also known as the Winkler Method. The azide modification of this method is commonly used since it is subject to fewer interferences.

Ion exchange resin, 57 Bead-like material that removes ions from water. Used in deionizers.

Iron bacteria, 96 Bacteria that use dissolved iron as an energy source. They can create serious problems in a water system since they form large masses that clog well screens, pumps, and other equipment.

Jackson Turbidity Unit, 107 A unit of TURBIDITY based on the amount of light from a candle flame that passes through a column of turbid water. This procedure has generally been replaced by the NEPHELOMETRIC method of turbidity measurement.

Jar test apparatus, 53 An automatic stirring machine equipped with three to six paddles and a variable-speed motor drive. Used to conduct the jar test in order to evaluate the coagulation, flocculation, and sedimentation processes.

Langelier Saturation Index, 83 A numerical index that indicates whether calcium carbonate will be deposited or dissolved in a distribution system. The index is also used to indicate the corrosivity of water.

MCL, 3 See MAXIMUM CONTAMINANT LEVEL.

Magnetic stirrer, 59 A device used for mixing chemical solutions in the laboratory.

Maximum contaminant level, 3 The maximum permissible level of a contaminant in water as specified in the regulations of the SAFE DRINKING WATER ACT.

Membrane filter, 53, 59 A filter made of cellulose acetate with a uniform, small pore size. Used for MICROBIOLOGICAL examination.

Membrane filter method, 73 A laboratory method used for COLIFORM testing. The procedure uses an ultra-thin filter with a uniform pore size smaller than bacteria—less than a micron. After water is forced through the filter, the filter is incubated. Bacterial colonies with a green-gold sheen indicate presence of COLIFORM BACTERIA.

Meter(s), 62 An instrument (usually electronic) used to measure water quality parameters such as pH.

Methyl orange, 82 An indicator used when measuring the total alkalinity of a water sample.

Microbiological, 71 Relating to microorganisms and their life processes.

Microscope, 66 A device used to magnify extremely small objects so that they can be seen and studied with the naked eye.

Milk dilution bottle, 45 See DILUTION BOTTLE.

Milligrams per litre (mg/L), 4 A unit of the concentration of water or wastewater constituent. It is 0.001 gram of the constituent in 1000 millilitres of water. In reporting the results of water and wastewater analyses it has replaced the unit formerly used commonly, parts per million, to which it is approximately equivalent.

Modified Winkler method, 92 A modification of the standard Winkler (IODOMETRIC) METHOD that uses an alkali-iodide-azide reagent to make the procedure less subject to interferences.

Mohr pipet, 47 A pipet with a graduated stem used to measure and transfer liquids when great accuracy is not required.

Monitoring, 9 Routine observation, sampling, and testing of water samples taken from different locations within a water system to determine water quality, efficiency of treatment processes, and compliance with regulations.

Mottled, 93 Spotted or blotched. Teeth can become mottled if excessive amounts of fluoride are consumed during the years of teeth formation.

Muffle furnace, 54 A high-temperature oven used to ignite and burn volatile solids. Usually operated at temperatures near 600°C.

Multiple-tube fermentation method, 73 A laboratory method used for COLIFORM testing, which uses a nutrient broth placed in culture tubes. Gas production indicates presence of COLIFORM BACTERIA.

NTU, 88 See NEPHELOMETRIC TURBIDITY UNIT.

National Interim Primary Drinking Water Regulations (NIPDWRs), 1 Regulations developed under the SAFE DRINKING WATER ACT. The NIPDWRs establish MCLs, monitoring requirements, and reporting procedures for contaminants in drinking water, which endanger human health.

Negative sample, 74 When referring to the MULTIPLE-TUBE FERMENTATION or MEMBRANE FILTER test, any sample that does not contain COLIFORM BACTERIA.

Nephelometer, 5 An instrument that measures TURBIDITY by measuring the amount of light scattered by turbidity in a water sample. It is the only instrument approved by the US Environmental Protection Agency to measure turbidity in treated drinking water.

Nephelometric turbidimeter, 65, 90, 107 See NEPHELOMETER.

Nephelometric turbidity unit (NTU), 5, 65, 88, 105 The amount of TURBIDITY in a water sample as measured using a NEPHELOMETER.

Nomographic method, 94 A method that uses a graph or other diagram to solve formulas and equations.

Non-community system, 3 As defined by the NIPDWRs, any water system serving a non-residential (transient) population such as tourists. A non-community system is a PUBLIC WATER SYSTEM that has at least 15 service connections used by non-residential consumers or serves a daily average of at least 25 non-residential consumers at least 60 days a year.

Organic chemical, 5 Chemical substance of animal or vegetable origin having carbon in its molecular structure.

Oven, 54 A chamber used to dry, burn, or sterilize materials.

Oxidant, 92 Any chemical substance that promotes oxidation. (See OXIDIZE.)

Oxidize, 99 To chemically combine with oxygen.

pH meter, 63 A sensitive voltmeter used to measure the pH of liquid samples.

pHs, 83 The theoretical pH at which calcium carbonate will neither dissolve nor precipitate. Used to calculate the Langelier Index.

Parts per million (ppm), 4 The number of weight or volume units of a constituent present with each one million units of the solution or mixture. Formerly used to express the results of most water and wastewater analyses but recently replaced by the ratio MILLIGRAMS PER LITRE. For drinking water analyses, concentrations in ppm and mg/L are equivalent.

Pathogens (pathogenic), 5, 71 Disease-causing organisms.

Petri dish, 46 A shallow glass or plastic dish with vertical sides, a flat bottom, and loose fitting cover. Used for growing MICROBIOLOGICAL cultures.

Phenanthroline method, 96 A COLORIMETRIC procedure used to determine the concentration of iron in water.

Phenolphthalein indicator, 94 A chemical INDICATOR (color-changing) used in several tests including tests for alkalinity, carbon dioxide, and pH.

Photometer, 62 An instrument used to measure the intensity of light transmitted through a sample or the degree of light absorbed by a sample.

Picocurie (pCi), 9 A measure of the disintegration of a particular RADIONUCLIDE.

Pipet, 47 Slender glass or plastic tube used to measure and transfer small volumes (usually less than 25 mL) of liquids.

Platinum-cobalt method, 91 A procedure used to determine the amount of color in water.

Positive sample, 74 When referring to the MULTIPLE-TUBE FERMENTATION or MEMBRANE FILTER test, any sample that contains COLIFORM BACTERIA.

Potentiometrically, 94 Any laboratory procedure that measures a difference in electric potential (voltage) to indicate the concentration of a constituent in water.

Precipitate, 85 To separate a substance from a solution or suspension by a chemical reaction.

Presumptive test, 73 The first major step in the MUTLIPLE-TUBE FERMENTATION test. The step presumes (indicates) presence of COLIFORM BACTERIA based on gas production in nutrient broth after INCUBATION.

Probe method, 98 See ELECTRODE METHOD.

Public water system, 3 As defined by the SAFE DRINKING WATER ACT, any system, publicly or privately owned, that serves at least 15 service connections 60 days out of the year or serves an average of 25 people at least 60 days out of the year.

Radionuclide, 9 A material with an unstable atomic nucleus, which spontaneously decays or disintegrates, producing radiation.

Reagent bottle, 48 A bottle made of borosilicate glass fitted with a ground glass stopper. Used to store reagents (standard chemical solutions).

Recarbonation, 94 The process of adding carbon dioxide as a final stage in the lime-soda ash softening process to convert carbonate to bicarbonates. This prevents precipitation of carbonates in the distribution system.

Refrigerator, 55 A cabinet used to store chemical solutions and preserve water samples.

Representative sample, 24 A sample that contains all the same constituents that are in the water from which it was taken.

Routine (required) sample, 9 A sample required by the NIPDWRs to be taken at regular intervals to determine compliance with the MCLs.

SPADNS method, 93 A COLORIMETRIC procedure used to determine the concentration of fluoride ion in water. SPADNS is the chemical reagent used in the test.

Safe Drinking Water Act, 1 A federal law enacted December 16, 1974 that sets up a cooperative program among local, state, and federal agencies to ensure safe drinking water for the consumer.

Sample bottle, 49 Wide-mouth glass or plastic bottle used for taking MICROBIOLOGICAL and chemical water samples.

Secondary Drinking Water Regulations, 1 Regulations developed under the SAFE DRINKING WATER ACT that establish maximum levels for substances affecting the taste, odor, or color (aesthetic characteristics) of drinking water.

Selective adsorption, 103 A method used in GAS CHROMATOGRAPHY to separate organic compounds so their concentrations can be determined.

Sequestering, 95 A chemical reaction in which certain chemicals (sequestering or chelating agents) "tie-up" other chemicals, particularly metal ions, so that the chemicals no longer react. Sequestering agents are used to prevent the formation of precipitates or other compounds.

Single-pan balance, 60 A BALANCE used to make quick, accurate weight measurements. The material to be weighed is placed on the pan and counterweights, located on arms (beams) beneath the pan, are adjusted to balance the material, thus indicating the weight. Also known as a beam balance.

Specific ion meter, 64 A sensitive voltmeter used to measure the concentration of specific ions, such as fluoride, in water. Electrodes designed specifically for each ion must be used.

Spectrophotometer, 63 A PHOTOMETER that uses a diffraction grating or a prism to control the light wavelengths used for specific analysis.

Splash goggles, 55 Safety goggles with shatter-proof lenses designed to provide a tight covering around the eyes, protecting them from chemicals and flying particles.

Stable, 96 Resistant to change.

Standard plate count, 71 A laboratory procedure for estimating the total bacterial count in a water sample. Also called the total plate count or total bacterial count.

Strip, 99 To remove gases from water by passing large volumes of air through the water.

TD, 47 A mark on a PIPET indicating "to deliver." The pipet is calibrated to deliver the calibrated volume of the pipet with a small drop left in the tip.

TON, 99 THRESHOLD ODOR NUMBER.

Tests, 81 Laboratory procedures used to determine the concentration of constituents in water.

Test tube, 49 Slender glass or plastic tube with an open top and rounded bottom. Used for a variety of tests.

Threshold odor number, 99 A number indicating the greatest dilution of a water sample (using odor-free water) that still yields a noticeable odor.

Time composite, 25 A COMPOSITE SAMPLE consisting of several equal-volume samples taken at specified times.

Titration, 44 A method of analyzing the composition of a solution by adding known amounts of a standardized solution until a given reaction or end point (color change, precipitation, or conductivity change) is produced.

Titrimetric method, 94 Any laboratory procedure that uses TITRATION to determine the concentration of a constituent in water.

Total coliform test, 71 Either the MULTIPLE-TUBE FERMENTATION or MEMBRANE FILTER test. Both tests indicate the presence of the entire coliform group or total COLIFORM.

Transect, 27 An imaginary line along which samples are taken at specified intervals. Transect sampling is usually done on large bodies of water such as rivers and lakes.

Transfer pipet, 47 See VOLUMETRIC PIPET.

Trihalomethanes (THMs), 5 A group of compounds formed when natural organic compounds from decaying vegetation and soil (such as humic and fulvic acids) react with chlorine.

Turbidimeter, 64 An instrument that measures the amount of light impeded or scattered by suspended particles in the water sample using a standard suspension as a reference.

Turbidity, 5 A physical characteristic of water making the water appear cloudy. The condition is caused by the presence of suspended matter.

Utility oven, 54 Laboratory oven used primarily to dry labware and chemicals prior to weighing or to sterilize labware.

Vacuum pump, 59 A pump used to provide a partial vacuum needed for filtering operations such as the MEMBRANE FILTER test.

Viable, 76 Capable of living.

Volumetric flask, 45 A squat bottle with a long, narrow neck. Used to prepare fixed volumes of solution. Each FLASK is CALIBRATED for a single volume only.

Volumetric pipet, 47 A PIPET CALIBRATED to deliver a single volume only.

Water still, 57 A device used to produce distilled water by evaporation and condensation of tap water.

Answers to Review Questions

Answers to Review Questions

Module 1 Drinking Water Standards

1. Skill to perform a variety of procedures, knowledge of the measures and principles used by biologists and chemists, understanding of the laws and regulations governing the production and distribution of drinking water.

2. Routine monitoring ensures that each component of the water system is in proper operating condition and checks the efficiency of various water treatment processes.

3. To develop national drinking water regulations.

4. To implement and monitor public water systems.

5. (1) A community system serving the same people year round, and (2) a non-community system serving intermittent users.

6. Community system.

7. (1) Inorganic chemicals, (2) Organic chemicals, (3) Turbidity, (4) Microbiological contaminants, and (5) Radiological contaminants.

8. Surface water, because it is of less uniform quality and more subject to contamination.

9. No, some systems may have stricter requirements.

10. State notification within seven days, check sampling within a specified time period (Table 1-6), mail notification, newspaper and broadcast notification.

11. Coliform was present in 6 out of 30 tubes (20 percent) during the month; this is a violation of the first part of the MCL, which specifies that coliform must not be present in more than 10 percent of the samples per month. (See *Basic Science Concepts and Applications*, Mathematics Section, Percent.)

    The system took six samples during the month, so the second part of the MCL is the regulation relating to systems taking less than 20 samples per month. That regulation states that not more than one monthly sample can have three or more portions positive. Since only one sample had three portions positive, the second part of the MCL was not violated.

    When the third sample taken during the month was found to be positive in three portions, the system should have begun check sampling and continued check sampling daily until no positive tubes were found two days in a row. There is no indication that check samples were taken, so the system has violated the regulations regarding check sampling.

12. Coliform was present in 4 of the 35 tubes (11 percent) during the month: this is a violation of the first part of the MCL, which specifies that coliform must not be present in more than 10 percent of the samples per month. Note that only the routine samples are counted for this determination. (See *Basic Science Concepts and Applications*, Mathematics Section, Percent.)

    The system took seven routine samples during the month, so the second part of the MCL is determined by the regulation relating to systems taking less than 20 samples per month. That regulation states that not more than one monthly sample can have three or more portions positive. Since only one sample had three portions positive, the second part of the MCL was not violated.

    When a sample was found to have three positive portions, check sampling was begun immediately and continued daily until two samples in a row had no positive tubes. This satisfies the requirements for check sampling.

13. Coliform was present at the average level of one colony (rounded from 1.27 colonies) per 100 mL for all routine samples taken during the month (11 samples taken, 14 colonies found). This is not a violation of the MCL, which states that the average level may not exceed one colony per 100 mL. Note that only the routine samples are counted for this determination. (See *Basic Science Concepts and Applications*, Mathematics Section, Rounding and Estimating, Averages.)

The system took 11 routine samples during the month, so the second part of the MCL is determined by the regulation relating to systems taking less than 20 samples per month. That regulation states that the coliform level must not exceed four colonies per month in more than one of the samples taken. Since only one of the routine samples had a colony count greater than four, the second part of the MCL was not violated.

When the routine sample was found to have five colonies per 100 mL, check sampling was begun immediately and continued until two samples in a row showed less than one colony per 100 mL. This satisfies the requirements for check sampling.

14. Population served.

15. Those affecting taste, odor, or appearance.

16. In instances where the taste, odor, or appearance of the public drinking water is unpleasant, consumers will often find other sources of water, which may be less safe.

Module 2 Sample Collection, Preservation, and Storage

1. Grab sample and composite sample.

2. A grab sample is a single volume of water taken all at one time from a single place; a composite sample is a mixture of grab samples taken at different times from the same place.

3. Requirements of the laboratory testing the sample.

4. 100 mL, 25 mL, 100 mL, 500 mL, 300 mL.

5. Raw-water transmission lines, ground water, streams and rivers, lakes and reservoirs.

6. Points should not be located immediately downstream from chemical additions; samples should be taken from main stream of flow.

7. Coliform bacteria, chlorine residual.

8. Chlorine residual.

9. Bacteriological problems and tastes and odors.

10. No. The faucet must run for 2 to 5 min to flush stagnant water from the service line.

11. One sample should be taken from the hot-water tap after the water has run and become hot, another should be taken from the cold-water tap.

12. Date of sample, time of sample, location sampled, type of sample, tests to be run, name of person sampling, preservatives used, bottle number.

13. Pesticides, radiochemicals.

14. Refrigeration, pH adjustment.

15. Stability of the constituent being tested for, whether or not the sample can be preserved.

Module 3 Water Laboratory Equipment and Instruments

1. (a) glass tubes, graduated over part of length, has stopcock; used to dispense solutions during titration; 10, 25, 50 mL; (b) tall, cylindrical container made of glass or plastic, usually has pour spout and hexagonal base; used to quickly measure liquids; 10 to 4000 mL; (c) shallow, vertical-sided dish with flat bottom and loose-fitting cover, made of glass or plastic, transparent; used for culturing standard plate counts and membrane filters; 100 mm × 15 mm or 50 mm × 12 mm; (d) square, glass bottles with narrow mouths, threaded to receive a screw cap; used to dilute bacteriological samples; 160-mL capacity; (e) hollow, slender, glass tube with rounded bottom and without flared lips; used for a variety of laboratory tests including multiple-tube fermentation tests and biochemical test for bacterial identification.

2. Name of chemical and chemical formula, concentration, date prepared, initials of person who prepared reagent, expiration date.

3. Detergent wash, acid wash with 10-percent HCl, hot tap water rinse, distilled water rinse.

4. Mechanical equipment of prime importance to a test or procedure.

5. (a) sealable container usually has a tight-fitting glass cover and ground-glass flanged closures; provides a place where heated items can cool slowly prior to weighing and provides a moisture-free environment so that items being cooled will not gain moisture weight before weighing; (b) consists of a filter holder base, a membrane filter, and filter funnel, fits on a vacuum filter flask or vacuum manifold; used for small particle filtration; (c) oven which operates in the temperature range from 30 to 350° C, gravity convection or forced air, a vacuum can be applied to some; used for drying samples prior to weighing or for sterilizing labware; (d) delivers a torrent of water in a uniform pattern to wash the body, has easy-to-grab pull chain or paddle, can be attached to an eye/face wash unit; used to rapidly cleanse the body of dangerous chemicals.

6. They have refrigeration units allowing them to achieve temperatures below room temperature.

7. Dissolved minerals, uncombined gases, all kinds of nonvolatile contaminants.

8. Danger to a laboratory water supply by cross connection.

9. Aid in filtration.

10. Filter paper, glass-fiber filters, membrane filters.

11. A pan balance, an analytical balance; respectively.

12. Over a period of time, the material being weighed may gain weight from moisture in the air.

13. Spectrophotometer. It provides selection from the complete spectrum of light.

Module 4 Microbiological Tests

1. Disease-causing organism.

2. Coliform bacteria are considered a good indicator of pathogens for the

following reasons: they are always present in contaminated water; they are always absent when contamination is not present; they survive longer in water than pathogens; they can be easily identified.

3. The production of gas within 48 hours.

4. The confirmed tests should be run on all positive tubes from the presumptive test.

5. Because coliform organisms are present in soil and plants, improper sampling can result in positive tests that are not due to system contamination.

6. The standard plate count indicates total population of bacteria in water. The test results determine effectiveness of treatment processes and general quality of water in the distribution system.

Module 5 Physical/Chemical Tests

1. Alkalinity must be present for effective coagulation with alum. Alkalinity must be determined to calculate the proper lime and soda ash needed for water softening. Alkalinity is a key factor in determining if water is corrosive.

2. Titration to an endpoint. The endpoint can be determined with a pH meter or color indicators.

3. Alkalinity, temperature, total dissolved solids, pH, and calcium.

4.
$$LI = pH - pH_s$$
$$pH_s = A + B - \log(Ca^{+2}) - \log(\text{alkalinity})$$
$$= 2.30 + 9.86 - 2.30 - 1.90$$
$$= 7.96$$
$$LI = 7.8 - 7.96$$
$$= -0.16$$

Water is unstable indicating that corrosion may be a problem. (See *Basic Science Concepts and Applications*, Chemistry Section, Chemistry of Treatment Processes [Scaling and Corrosion Control]).

5. The Langelier Index is only an indicator of stability. It is not an exact measure of corrosion or deposition tendencies of a water.

6. Free and combined.

7. Total chlorine added − free available (or combined) chlorine after a specified time.

8. 8 mg/L−0.5 mg/L = 7.5 mg/L (See *Basic Science Concepts and Applications,* Chemistry Section, Dosage Problems [Chlorine Dosage/Demand/Residual Calculations]).

9. 3 mg/L − 2.6 mg/L = 0.4 mg/L

10. Free available chlorine residual.

11. DPD (colorimetric) and amperometric titration (electrometric).

12. To determine the optimum dosage of coagulant.

13. a. $0.10/lb × 250 lb/mil gal = $25.00/mil gal.
    b. $0.20/lb × 176 mil gal = $35.20/mil gal.
    c. Aluminum sulfate. $35.20 − $25.00 = $10.20/mil gal.

14. Most color problems originate from natural organic compounds in soil and vegetation. Surface water usually contains more of these compounds due to runoff.

15. Dissolved oxygen can be used to oxidize certain constituents in water such as iron and manganese.

16. The electrode method since it is not as subject to interferences. It is also easier and faster to use than the Winkler method.

17. To help prevent cavities in children.

18. Daily.

19. Corrosion.

20. pH, bicarbonate alkalinity, temperature, and total dissolved solids.

21. Hardness is a measure of calcium and magnesium salts in water.

22. Titration with EDTA. Sequestration.

23. Staining of plumbing fixtures and laundry, precipitates in mains.

24. Corrosion of cast-iron mains.

25. pH is a measure of the intensity of the acid or alkaline condition of water.

26. Acidic—0 to 6.9; neutral—7.0; alkaline—7.1 to 14.

27. The electrode, or electrometric, method.

28. The sense of smell is much more sensitive than the sense of taste.

29. Odor tests may be used to evaluate the taste and odor removal capabilities of a water plant and the efficiency of activated carbon, chlorine dioxide, ozone, or potassium permanganate. Odor tests also help detect nuisance-causing tastes and odors.

30. Glass.

31. Most chemical reactions slow down and become less efficient as temperature is lowered. As a result, increased mixing or detention time may be required.

32. A gravimetric procedure whereby a sample is evaporated and the residue is weighed.

33. Compounds formed by the reaction of natural organics such as humic acid with chlorine.

34. A reducing agent such as sodium thiosulfate or sodium sulfite.

35. 24 hours in the dark.

36. Nephelometric method.

37. No. Suspended solids refer to the concentration of solids in suspension. Turbidity is a measurement of how much light is scattered by the solids.

38. Turbidity can provide shelter for microorganisms and this hinders disinfection. A turbid water even when disinfected has a greater potential for containing live pathogens than clear water.

39. No. Coagulation/flocculation and settling should reduce sedimentation turbidity to the range of 1 to 10 NTU. Filtration should reduce turbidity to less than 1 NTU.

40. Atomic absorption spectrophotometer.